SpringerBriefs in Mathematics

SpringerBriefs in Mathematics showcases expositions in all areas of mathematics and applied mathematics. Manuscripts presenting new results or a single new result in a classical field, new field, or an emerging topic, applications, or bridges between new results and already published works, are encouraged. The series is intended for mathematicians and applied mathematicians.

BCAM SpringerBriefs

BCAM SpringerBriefs aims to publish contributions in the following disciplines: Applied Mathematics, Finance, Statistics and Computer Science. BCAM has appointed an Editorial Board, who evaluate and review proposals.

Typical topics include: a timely report of state-of-the-art analytical techniques, bridge between new research results published in journal articles and a contextual literature review, a snapshot of a hot or emerging topic, a presentation of core concepts that students must understand in order to make independent contributions.

Please submit your proposal to the Editorial Board or to Francesca Bonadei, Executive Editor Mathematics, Statistics, and Engineering: francesca.bonadei@springer.com.

basque center for applied **mathematics**

More information about this series at http://www.springer.com/series/10030

Jesús Martínez-Frutos · Francisco Periago Esparza

Optimal Control of PDEs under Uncertainty

An Introduction with Application to Optimal Shape Design of Structures

Jesús Martínez-Frutos
Computational Mechanics and Scientific
    Computing Group, Department
    of Structures and Construction
Technical University of Cartagena
Cartagena, Murcia, Spain

Francisco Periago Esparza
Department of Applied Mathematics
    and Statistics
Technical University of Cartagena
Cartagena, Murcia, Spain

ISSN 2191-8198   ISSN 2191-8201   (electronic)
SpringerBriefs in Mathematics
ISBN 978-3-319-98209-0   ISBN 978-3-319-98210-6   (eBook)
https://doi.org/10.1007/978-3-319-98210-6

Library of Congress Control Number: 2018951210

This Springer imprint is published by the registered company Springer Nature Switzerland AG
The registered company address is: Gewerbestrasse 11, 6330 Cham, Switzerland

To our families

Preface

Both the theory and the numerical resolution of optimal control problems of systems governed by—*deterministic*—partial differential equations (PDEs) are very well established, and several textbooks providing an introduction to the fundamental concepts of the mathematical theory are available in the literature [11, 16].

All of these books as well as most of the researches done in this field assume that the model's input data, e.g. coefficient in the principal part of the differential operator, right-hand side term of the equation, boundary conditions or location of controls, among others, are perfectly known. However, this assumption is unrealistic from the point of view of applications. In practice, input data of physical models may only be estimated [5]. In addition, as will be illustrated in this book, solutions of optimal control problems may exhibit a dramatic dependence on those uncertain data. Consequently, more realistic optimal control problems should account for uncertainties in their mathematical formulations.

Uncertainty appears in many other optimization problems. For instance, in structural optimization, where one aims to design the shape of a structure that minimizes (or maximizes) a given criterion (rigidity, stress, etc.). All physical parameters of the model, e.g. the loads, which are applied to the structure, are supposed to be known in advance. However, it is clear that these loads either vary in time or vary from one sample to another or are simply unpredictable.

Among others, the following two approaches are used to account for uncertainties in control problems:

- On the one hand, if there is no a priori information on the uncertain inputs short of lower and upper bounds on their magnitudes, then one may use a worst-case scenario analysis, which is formulated as a min-max optimization problem. More precisely, the idea is to minimize the maximal value of the cost function over a given set of admissible perturbations. The advantage of this formulation is that, in general, it does not require an exhaustive previous analysis of the uncertain parameters of the model, just the lower and upper bounds of those. However, this *a priori* advantage may become a disadvantage because if the upper and lower bounds on the uncertain data are relatively large, then optimal

solutions obtained from this approach may be too conservative, leading to a
poor performance of the physical system under consideration [1].

- On the other hand, when statistical information on the random inputs is available, it is natural to model uncertainties using probabilistic tools. In some
applications, the mean and the spatial correlation of these inputs are estimated
from experimental data. From this, assuming a Gaussian-type distribution of
such inputs, those may be modelled by using, for example, Karhunen–
Loève-type expansions. More elaborated statistical methods may also be used to
represent input parameters [5, 15]. Due to the great interest in applications, very
efficient computational methods for solving PDEs with random inputs have been
developed during the last two decades (see, e.g. [2, 3, 6, 9]). These continuous
theoretical and numerical improvements in the analysis and resolution of random PDEs have enabled the development of control formulations within a
probabilistic setting (see, e.g. [7, 13, 14]). Probably, the fact that this probabilistic framework is fed with experimental data is the main drawback of this
approach. On the other hand, optimal solutions obtained from these probabilistic
methods are, in general, better than those obtained from a worst-case scenario.

This book aims at introducing graduate students and researchers to the basic
theoretical and numerical tools which are being used in the analysis and numerical
resolution of optimal control problems constrained by PDEs with random inputs.
Thus, a probabilistic description of those inputs is assumed to be known in advance.
The text is confined to facilitating the transition from control of deterministic PDEs
to control of PDEs with random inputs. Hence, a previous elementary knowledge of
optimal control of deterministic PDEs and probability theory is assumed. We
recommend the monograph [16] for a basic and very complete introduction to
deterministic optimal control. Concerning probability theory, any introductory
textbook on probabilities, e.g. [8], provides the bases which are required in this
book. See also [12, Chap. 4] for a complete review of this material.

The book is composed of seven chapters:

Chapter 1 is devoted to a very brief introduction to uncertainty modelling, by use
of random variables and random fields, in optimal control problems. To this end,
three very simple but illustrative examples are introduced. These examples cover
the cases of elliptic and evolution (first and second order in time) PDEs. We will
consistently use these models, but especially the elliptic one, to illustrate (some of)
the main methods in optimal control of random PDEs.

Chapter 2 introduces the mathematical and numerical tools which are required in
the remaining chapters. Special attention is paid to the numerical approximation of
random fields by means of Karhunen–Loève (KL) expansions.

Chapter 3 focusses on the mathematical analysis of optimal control problems for
random PDEs. Firstly, well-posedness of such random PDEs is proved by
extending the classical methods for deterministic PDEs (Lax–Milgram theorem for
the elliptic case and the Galerkin method for evolution equations) to a suitable
probabilistic functional framework. Secondly, the existence of solutions for robust
and risk-averse optimal control problems is proved. The former formulation

naturally arises when one is primarily interested in minimizing statistical quantities of the cost functions such as its expectation. In order to reduce the dispersion of the objective function around its mean, it is reasonable to consider measures of statistical dispersion such as standard deviation. This choice, usually referred to in the literature as *robust control* [10], aims at obtaining optimal solutions less sensitive to variations in the input data. However, if one aims to obtain optimal controls for unlike realizations of the random input data (which therefore have a very little impact on the mean and variance of the objective function but which could have catastrophic consequences), then risk-averse models ought to be considered. In this sense, *risk-averse control* [4] minimizes a risk function that quantifies the expected loss related to the damages caused by catastrophic failures. A typical example which illustrates the main difference between robust and risk-averse control is the following: imagine that our optimization problems amount to finding the optimal shape of a bridge. If one is interested in maximizing the rigidity of the bridge in the face of daily random loads (e.g. cars and people passing across the bridge), then a robust formulation of the problem is a suitable choice. However, if the interest is to design the bridge to be able to withstand the effects of an earthquake, then one must use a risk-averse approach. In probabilistic terms, robust control focusses on the first two statistical moments of the distribution to be controlled. Risk-averse control also takes into account the tail of such distribution.

Chapter 4 is concerned with numerical methods for robust optimal control problems. Both gradient-based methods and methods based on the resolution of the first-order optimality conditions are described. Smoothness with respect to the random parameters is a key ingredient in the problems considered in this chapter. Thus, we focus on stochastic collocation (SC) and stochastic Galerkin (SG) methods [2], which, in this context, are computationally more efficient than the classical brute force Monte Carlo (MC) method. An algorithmic exposition of these methods, focussing on the practical numerical implementation, is presented. To this end, MATLAB computer codes for the examples studied in this chapter are provided in http://www.upct.es/mc3/en/book.

Chapter 5 deals with the numerical resolution of risk-averse optimal control problems. An adaptive gradient-based minimization algorithm is proposed for the numerical approximation of this type of problem. An additional difficulty which emerges when dealing with risk-averse-type cost functionals is the lack of smoothness with respect to the random parameters. A combination of a stochastic Galerkin method with a Monte Carlo sampling method is used for the numerical approximation of the cost functional and of its sensitivity. As in the preceding chapter, an algorithmic approach supported with MATLAB computer codes for several illustrative examples is presented.

Chapter 6 aims to illustrate how the methods presented in the book may be used to solve optimization problems of interest in real applications. To this end, robust and risk-averse formulations of the structural optimization problem are analysed from theoretical and numerical viewpoints.

Finally, Chap. 7 includes some concluding remarks and open problems.

A list of references is included in each chapter. Also, links are provided to websites hosting open-source codes related to the contents of the book.

We have done our best in writing this book and very much hope that it will be useful for readers. This would be our best reward. However, since the book is concerned with uncertainties, very likely there are uncertainties in the text associated with inaccuracies or simply errors. We would be very grateful to readers who communicate any such uncertainties to us. A list of possible errata and improvements will be updated on the website: http://www.upct.es/mc3/en/book/.

We would like to thank Enrique Zuazua for encouraging us to write this book. We also acknowledge the support of Francesca Bonadei, from Springer, and of the anonymous reviewers whose comments on the preliminary draft of the book led to a better presentation of the material. Special thanks go to our closest collaborators (David Herrero, Mathieu Kessler, Francisco Javier Marín and Arnaud Münch) during these past years. Working with them on the topics covered by this text has been a very fruitful and enriching experience for us. Finally, we acknowledge the financial support of Fundación Séneca (Agencia de Ciencia y Tecnología de la Región de Murcia, Spain) under project 19274/PI/14 and the Spanish Ministerio de Economía y Competitividad, under projects DPI2016-77538-R and MTM2017-83740-P.

Cartagena, Spain   Jesús Martínez-Frutos
June 2018   Francisco Periago Esparza

References

1. Allaire, A., Dapogny, Ch.: A deterministic approximation method in shape optimization under random uncertainties. SMAI J. Comp. Math. **1**, 83–143 (2015)
2. Babuška, I., Novile, F., Tempone, R.: A Stochastic collocation method for elliptic partial differential equations with random input data. SIAM Rev. **52** (2), 317–355 (2010)
3. Cohen, A., DeVore, R.: Approximation of high-dimensional parametric PDEs. Acta Numer. Graduate Studies in Mathematics, 1–159 (2015)
4. Conti, S., Held, H., Pach, M., Rumpf, M., Schultz, R.: Risk averse shape optimization. SIAM J. Control Optim. **49**(3), 927–947 (2011)
5. de Rocquigny, E., Devictor, N., Tarantola, S.: Uncertainty in industrial practice: a guide to quantitative uncertainty management. Wiley, New Jersey (2008)
6. Gunzburger, M. D., Webster, C., Zhang, G.: Stochastic finite element methods for partial differential equations with random input data. Acta Numer. **23**, 521–650 (2014)
7. Gunzburger, M. D., Lee, H-C., Lee, J.: Error estimates of stochastic optimal Neumann boundary control problems. SIAM J. Numer. Anal. **49** (4), 1532–1552 (2011)
8. Jacod, J., Protter, P.: Probability essentials (Second Edition). Springer, Heidelberg (2003)
9. Le Maître, O. P., Knio, O. M.: Spectral methods for uncertainty quantification. with applications to computational fluid dynamics. Springer, New York (2010)
10. Lelievre, N., Beaurepaire, P., Mattrand, C., Gayton, N., Otsmane, A.: On the consideration of uncertainty in design: optimization—reliability—robustness. Struct. Multidisc. Optim. (2016) https://doi.org/10.1007/s00158-016-1556-5

11. Lions, J. L.: Optimal control of systems governed by partial differential equations. Springer, Berlin (1971)
12. Lord, G. L., Powell, C. E., Shardlow, T.: An introduction to computational stochastic PDEs. Cambridge University Press, Cambridge (2014)
13. Lü, Q, Zuazua, E.: Averaged controllability for random evolution partial differential equations. J. Math. Pures Appl. **105**(3), 367–414 (2016)
14. Martínez-Frutos, J., Herrero-Pérez, D., Kessler, M., Periago, F.: Robust shape optimization of continuous structures via the level set method. Comput. Methods Appl. Mech. Engrg. **305**, 271–291 (2016)
15. Smith, R. C. Uncertainty quantification. Theory, implementation, and applications. Computational Science & Engineering, Vol. 12. Society for Industrial and Applied Mathematics (SIAM), Philadelphia, PA (2014)
16. Tröltzsch, F.: Optimal control of partial differential equations: Theory, methods and applications. Graduate Studies in Mathematics Vol. 112. AMS. Providence, Rhode Island (2010)

Contents

About the Authors

Jesús Martínez-Frutos obtained his Ph.D. from the Technical University of Cartagena, Spain, in 2014 and is currently Assistant Professor in the Department of Structures and Construction and a member of the Computational Mechanics and Scientific Computing group at the Technical University of Cartagena, Spain.

His research interests are in the fields of robust optimal control, structural optimization under uncertainty, efficient methods for high-dimensional uncertainty propagation and high-performance computing using GPUs, with special focus on industrial applications. Further information is available at: http://www.upct.es/mc3/en/dr-jesus-martinez-frutos/.

Francisco Periago Esparza completed his Ph.D. at the University of Valencia, Spain, in 1999. He is currently Associate Professor in the Department of Applied Mathematics and Statistics and a member of the Computational Mechanics and Scientific Computing group at the Technical University of Cartagena, Spain.

His main research interests include optimal control, optimal design and controllability. During his research trajectory, he has published a number of articles addressing not only theoretical advances of these problems but also their application to real-world engineering problems. During recent years, his research has focussed on optimal control for random PDEs. Further information is available at: http://www.upct.es/mc3/en/dr-francisco-periago-esparza/.

Acronyms and Initialisms

a.e.	Almost everywhere
a.s.	Almost sure
ANOVA	Analysis of Variance
ATD	Anisotropic Total Degree
e.g.	for example (exempli gratia)
etc.	etcetera
FEM	Finite Element Method
HDMR	High-Dimensional Model Representation
i.e.	that is (id est)
i.i.d.	independent and identically distributed
KL	Karhunen-Loève
MC	Monte Carlo
PC	Polynomial Chaos
PDEs	Partial Differential Equations
PDF	Probability Density Function
POD	Proper Orthogonal Decomposition
r.h.s.	right-hand side
RB	Reduced Basis
RPDEs	Random Partial Differential Equations
SC	Stochastic Collocation
SG	Stochastic Galerkin
UQ	Uncertainty Quantification
w.r.t.	With respect to

Abstract

This book offers a direct and comprehensive introduction to the basic theoretical and numerical concepts in the emergent field of optimal control of partial differential equations (PDEs) under uncertainty. The main objective of the book is to provide graduate students and researchers with a smooth transition from optimal control of *deterministic* PDEs to optimal control of *random* PDEs. Coverage includes uncertainty modelling in control problems, variational formulation of PDEs with random inputs, robust and risk-averse formulations of optimal control problems, existence theory and numerical resolution methods. The exposition is focussed on running the whole path starting from uncertainty modelling and ending in the practical implementation of numerical schemes for the numerical approximation of the considered problems. To this end, a selected number of illustrative examples are analysed in detail along the book. Computer codes, written in MATLAB, for all these examples are provided. Finally, the methods presented in the text are applied to the mathematical analysis and numerical resolution of the, very relevant in real-world applications, structural optimization problem.

Keywords Uncertainty quantification · Partial differential equations with random inputs · Stochastic expansion methods · Robust optimal control
Risk-averse optimization · Structural shape optimization under uncertainty

Chapter 1
Introduction

... in the small number of things which we are able to know with certainty, even in the mathematical sciences themselves, the principal means for ascertaining—induction and analogy—are based on probabilities; so that the entire system of human knowledge is connected with the theory set forth in this essay..

Pierre Simon de Laplace.
A Philosophical Essay on Probability, 1816.

1.1 Motivation

Mathematical models (and therefore numerical simulation results obtained from them) of physical, biological and economical systems always involve errors which lead to discrepancies between simulation results and the results of real systems. In particular, simplifications of the mathematical model are often performed to facilitate its resolution but these generate *model errors*, the numerical resolution method introduces *numerical errors*, and a limited knowledge of the system's parameters, such as its geometry, initial and/or boundary conditions, external forces and material properties (diffusion coefficients, elasticity modulus, etc.), induces additional errors, called *data errors*. Some of the above errors may be reduced, e.g., by building more accurate models, by improving numerical resolution methods, by additional measurements, etc. This type of errors or uncertainties that can be reduced is usually called *epistemic or systematic uncertainty*. However, there are other sources of randomness that are intrinsic to the system itself and hence cannot be reduced. It is the so-called *aleatoric or statistical uncertainty*. A classical example of this type is Uncertainty Principle of Quantum Mechanics.

© The Author(s), under exclusive license to Springer Nature Switzerland AG 2018
J. Martínez-Frutos and F. Periago Esparza, *Optimal Control of PDEs under Uncertainty*,
SpringerBriefs in Mathematics, https://doi.org/10.1007/978-3-319-98210-6_1

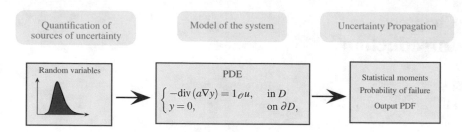

Fig. 1.1 Uncertainty Quantification in PDEs-based mathematical models

For the above reasons, it is clear that if one aims at obtaining more reliable numerical predictions from mathematical models, then these should account for uncertainty. Aleatoric uncertainty is naturally defined in a probabilistic framework. In principle, the use of probabilistic tools to model epistemic uncertainty is not a natural choice. However, there is a tendency in Uncertainty Quantification (UQ) to reformulate epistemic uncertainty as aleatoric uncertainty where probabilistic analysis is applicable [2]. In addition, from the point of view of numerical methods that will be proposed in this text, there is no difference between the above two types of uncertainty. Hence, a probabilistic framework for modelling and analysis of uncertainties is adopted throughout the book.

The term *Uncertainty Quantification* is broadly used to define a variety of methodologies. In [5, p. 1] it is defined as "the science of identifying, quantifying, and reducing uncertainties associated with models, numerical algorithms, experiments, and predicted outcomes or quantities of interest". Following [4, p. 525], in this text the term Uncertainty Quantification is used to describe "the task of determining information about the uncertainty in an output of interest that depends on the solution of a PDE, given information about the uncertainties in the system inputs". This idea is illustrated in Fig. 1.1.

Remark 1.1 When statistical information on uncertainties is not available, other techniques, such as worst case scenario analysis or fuzzy logic, may be used to deal with uncertainty. These tools are outside the scope of this text.

1.2   Modelling Uncertainty in the Input Data. Illustrative Examples

Among the different sources of uncertainty described in the preceding section, this text is focussed on optimal control problems whose state variable is a solution of a Random Partial Differential Equation (RPDE). The term *random* means that the PDE's input data (coefficients, forcing terms, boundary conditions, etc.) are uncertain and are described as random variables or random fields.

In the following, we present several illustrative examples that will be analysed in detail along the book.

1.2.1 The Laplace-Poisson Equation

Let $D \subset \mathbb{R}^d$, $d = 1, 2$ or 3 in applications, be a bounded domain and consider the following control system for Laplace-Poisson's equation

$$\begin{cases} -\text{div}\,(a\nabla y) = 1_{\mathcal{O}}u, & \text{in } D \\ y = 0, & \text{on } \partial D, \end{cases} \tag{1.1}$$

where $y = y(x)$, with $x = (x_1, \ldots, x_d) \in D$, is the state variable and $u = u(x)$ is the control function, which acts on the spatial region $\mathcal{O} \subset D$. As usual, $1_{\mathcal{O}}$ stands for the characteristic function of $\mathcal{O}$, i.e.,

$$1_{\mathcal{O}}(x) = \begin{cases} 1, & x \in \mathcal{O} \\ 0, & \text{otherwise.} \end{cases}$$

Here and throughout the text, both the divergence (div) and the gradient (∇) operators involve only derivatives with respect to the spatial variable $x \in D$. Recall that

$$\nabla y = \left(\frac{\partial y}{\partial x_1}, \ldots, \frac{\partial y}{\partial x_d} \right)$$

and

$$\text{div}\,(a\nabla y) = \sum_{j=1}^{d} \frac{\partial}{\partial x_j} \left(a \frac{\partial y}{\partial x_j} \right).$$

Equation (1.1) is a model for a variety of physical phenomena (deformation of elastic membranes, steady-state heat conduction, electrostatics, steady-state groundwater flow, etc.). For instance, in the case of heat conduction, y denotes the temperature of a solid which occupies the spatial domain D, a is the material thermal conductivity, and u represents an internal heat source, which heats/cools the body (e.g., by electromagnetic induction). For some nonhomogeneous materials, such as functionally graded materials, large random spatial variabilities in the material properties have been observed [1].

Another example is the modelling of a steady-state, single-phase fluid flow. In this case y represents the hydraulic head, u is a source/sink term, and a is the saturated hydraulic conductivity. In most aquifers, a is highly variable and never perfectly known (see [6, Chap. 1] where some experimental data for a are reported).

It is then natural to consider the coefficient a as a random function $a = a\,(x, \boldsymbol{\omega})$, where $x \in D$ and $\boldsymbol{\omega}$ denotes an elementary random event. As a consequence, the temperature y, solution of (1.1), becomes a random space function $y = y\,(x, \boldsymbol{\omega})$. Hereafter, the probability distribution of a is assumed to be known and thus an underlying abstract probability space $(\Omega, \mathscr{F}, \mathbb{P})$ is given. The sample space Ω is the set of all possible outcomes (e.g., all possible measures of thermal conductivity),

$\mathscr{F}$ is the σ—algebra of events (hence, subsets of Ω to which probabilities may be assigned), and $\mathbb{P} : \mathscr{F} \to [0, 1]$ is a probability measure.

A typical optimal control problem that arises in this context amounts to finding a control $u = u(x)$ such that its associated temperature $y = y(u)$ is the best approximation (in a least-square sense) to a desired stationary temperature $y_d(x)$ in D. Hence, the cost functional

$$\frac{1}{2} \int_D |y(x, \omega) - y_d(x)|^2 \, dx + \frac{\gamma}{2} \int_{\mathscr{O}} u^2(x) \, dx, \tag{1.2}$$

with $\gamma \geq 0$, is introduced. The second term in (1.2) is a measure of the energy cost needed to implement the control u. It is observed that (1.2) depends on the realization $\omega \in \Omega$ and hence a control u that minimizes (1.2) also depends on ω. Of course, one is typically interested in a control u, independent of ω, which minimizes (in some sense) the distance between $y(x, \omega)$ and $y_d(x)$. At first glance, it is natural to consider the averaged distance, with respect to ω, of the functional (1.2). The problem then is formulated as:

$$\begin{cases} \text{Minimize in } u : J(u) = \frac{1}{2} \int_{\Omega} \int_D |y(x, \omega) - y_d(x)|^2 \, dx \, d\mathbb{P}(\omega) + \frac{\gamma}{2} \int_{\mathscr{O}} u^2(x) \, dx \\ \text{subject to} \\ \qquad -\text{div}(a(x, \omega) \nabla y(x, \omega)) = 1_{\mathscr{O}} u(x), \quad (x, \omega) \in D \times \Omega \\ \qquad y(x, \omega) = 0, \qquad\qquad\qquad\qquad (x, \omega) \in \partial D \times \Omega \\ \qquad u \in L^2(\mathscr{O}). \end{cases}$$

$$\tag{1.3}$$

Following the usual terminology in Control Theory, throughout this text $y_d = y_d(x)$ is the so-called target function, $y = y(x, \omega)$ is the state variable and $u = u(x)$ the control. To avoid awful notations, we shall make explicit only the variables of interest that different functions depend on. For instance, to make explicit the dependence of the state variable y on the control u we will write $y = y(u)$.

1.2.2 The Heat Equation

Consider the following control system for the transient heat equation

$$\begin{cases} y_t - \text{div}(a \nabla y) = 0, & \text{in } (0, T) \times D \times \Omega \\ a \nabla y \cdot n = 0, & \text{on } (0, T) \times \partial D_0 \times \Omega \\ a \nabla y \cdot n = \alpha(u - y), & \text{on } (0, T) \times \partial D_1 \times \Omega, \\ y(0, x, \omega) = y^0(x, \omega) & \text{in } D \times \Omega, \end{cases} \tag{1.4}$$

where the boundary of the spatial domain $D \subset \mathbb{R}^d$ is decomposed into two disjoint parts $\partial D = \partial D_0 \cup \partial D_1$, and n is the unit outward normal vector to ∂D. In system (1.4), y_t denotes the partial derivative of y w.r.t. $t \in (0, T)$, and the control function $u = u(t, x)$ is the outside temperature, which is applied on the boundary region ∂D_1.

As reported in [1], in addition to randomness in the thermal conductivity coefficient $a = a(x, \omega)$, the initial temperature y^0 and the convective heat transfer coefficient α are very difficult to measure in practice and hence both are affected of a certain amount of uncertainty, i.e., $y^0 = y^0(x, \omega)$ and $\alpha = \alpha(x, \omega)$. In real applications α may also depend on the time variable $t \in [0, T]$ but, for the sake of simplicity, here it is assumed to be stationary.

As in the preceding example, the state variable $y = y(t, x, \omega)$ of system (1.4), which represents the temperature of the material point $x \in D$ at time $t \in [0, T]$ for the random event $\omega \in \Omega$, is a random function.

Assume that a desired temperature $y_d(x)$ is given and that it is aimed to choose the control u which must be applied to ∂D_1 in order to be as closer as possible to y_d at time T. Similarly to problem (1.3), one may consider the averaged distance between $y(T, x, \omega)$ and $y_d(x)$ as the cost functional to be minimized. Another possibility is to minimize the distance between the mean temperature of the body occupying the region D and the target temperature y_d, i.e.,

$$\int_D \left(\int_\Omega y(T, x, \omega)\, d\mathbb{P}(\omega) - y_d(x) \right)^2 dx.$$

Since only the mean of $y(T)$ is considered, there is no control on the dispersion of $y(T)$. Consequently, if the dispersion of $y(T)$ is large, then minimizing the expectation of $y(T)$ is useless because for a specific random event ω, the probability of $y(T, \omega)$ of being close to its average is small. It is then convenient to minimize not only the expectation but also a measure of dispersion such as the variance. Thus, the optimal control problem reads as:

$$\begin{cases} \text{Minimize in } u : J(u) = \int_D \left(\int_\Omega y(T, x, \omega)\, d\mathbb{P}(\omega) - y_d(x) \right)^2 dx + \frac{\gamma}{2} \int_D \text{Var}(y(T, x))\, dx \\ \text{subject to} \\ \qquad y = y(u) \quad \text{solves (1.4)}, \quad u \in L^2\left(0, T; L^2(\partial D_1)\right), \end{cases}$$

$$(1.5)$$

where $\gamma \geq 0$ is a weighting parameter, and

$$\text{Var}(y(T, x)) = \int_\Omega y^2(T, x, \omega)\, d\mathbb{P}(\omega) - \left(\int_\Omega y(T, x, \omega)\, d\mathbb{P}(\omega) \right)^2$$

is the variance of $y(T, x, \cdot)$.

1.2.3 The Bernoulli-Euler Beam Equation

Under Bernoulli-Euler hypotheses, the small random vibrations of a thin, uniform, hinged beam of length L, driven by a piezoelectric actuator located along the random interval $(x_0(\omega), x_1(\omega)) \subset D := (0, L)$ are described by the system:

$$
\begin{cases}
y_{tt} + (ay_{xx})_{xx} = v \left[\delta_{x_1(\omega)} - \delta_{x_0(\omega)} \right]_x, & \text{in } (0, T) \times D \times \Omega \\
y(t, 0, \omega) = y_{xx}(t, 0, \omega) = 0, & \text{on } (0, T) \times \Omega \\
y(t, L, \omega) = y_{xx}(t, L, \omega) = 0, & \text{on } (0, T) \times \Omega \\
y(0, x, \omega) = y^0(x, \omega), \quad y_t(0, x, \omega) = y^1(x, \omega), & \text{in } D \times \Omega,
\end{cases} \tag{1.6}
$$

where the beam flexural stiffness $a = EI$ is assumed to depend on both $x \in D$ and $\omega \in \Omega$. i.e. $a = a(x, \omega)$. As usual E denotes Young's modulus and I is the area moment of inertia of the beam's cross-section. Randomness in a is associated to a number of random phenomena, e.g. variabilities of conditions during the manufacturing process of the material's beam (see [3] and the references therein).

In system (1.6), $x_0(\omega)$ and $x_1(\omega)$ stand for the end points of the actuator. The dependence of these two points on a random event ω indicates that there is some uncertainty in the location of the actuator. $\delta_{x_j} = \delta_{x_j}(x)$ is the Dirac mass at the points $x_j \in D$, $j = 0, 1$.

The function $v : (0, T) \times \Omega \to \mathbb{R}$ represents the time variation of a voltage which is applied to the actuator, and which is assumed to be affected by some random perturbation. Since physical controller devices are affected of uncertainty, it is realistic to decompose the control variable into an unknown deterministic and a known stochastic components. Moreover, it is reasonable to assume that the stochastic part be modulated by the deterministic one. Thus, the function $v = v(t, \omega)$, which appears in (1.6), takes the form

$$
v(t, \omega) = u(t) \left(1 + \hat{u}(\omega) \right), \tag{1.7}
$$

where $u : (0, T) \to \mathbb{R}$ is the *(unknown)* deterministic control and $\hat{u}$ is a *(known)* zero-mean random variable which accounts for uncertainty in the controller device.

The random output $y(t, x, \omega)$ represents vertical displacement at time t of the particle on the centreline occupying position x in the equilibrium position $y = 0$. In the above equations, y_t and y_{tt} denote first and second derivatives w.r.t. $t \in (0, T)$, and the subscripts "$_x$" and "$_{xx}$" are first and second derivatives w.r.t. $x \in D$, respectively. See Fig. 1.2 for the problem configuration.

Let $H = L^2(D)$ be the space of (Lebesgue) measurable functions $v : D \to \mathbb{R}$ whose norm

Fig. 1.2 Problem configuration for the Bernoulli-Euler beam

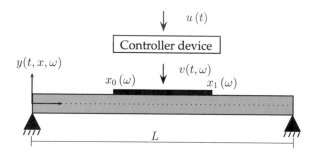

$$\|v\|_{L^2(D)} := \left(\int_D v^2(x)\, dx \right)^{1/2}$$

is finite. Consider the Sobolev space

$$V = H^2(D) \cap H_0^1(D) := \left\{ v : D \to \mathbb{R} : v, v', v'' \in L^2(D), \quad v(0) = v(L) = 0 \right\},$$

endowed with the norm

$$\|v\|_V = \|v\|_{L^2(D)} + \|v'\|_{L^2(D)} + \|v''\|_{L^2(D)}.$$

The topological dual space of V is denoted by V^*. For each random event $\omega \in \Omega$ and each control $u \in L^2(0, T)$, a measure of the amplitude and velocity of the beam vibrations at a control time $T > 0$ is expressed by

$$J(u, \omega) = \frac{1}{2} \left(\|y(T, \omega)\|_H^2 + \|y_t(T, \omega)\|_{V^*}^2 \right). \tag{1.8}$$

In this problem, one may be interested in computing a control u for rare occurrences of the random inputs (which therefore have a very little impact on the mean and variance of J), but which could have catastrophic consequences. For that purpose, the following risk averse model may be considered:

$$J_\varepsilon(u) = \mathbb{P}\{\omega \in \Omega : J(u, \omega) > \varepsilon\}, \tag{1.9}$$

where $\varepsilon > 0$ is a prescribed threshold parameter. The corresponding control problem is then formulated as:

$$\begin{cases} \text{Minimize in } u : J_\varepsilon(u) = \mathbb{P}\{\omega \in \Omega : J(u, \omega) > \varepsilon\} \\ \text{subject to} \\ \quad y = y(u) \quad \text{solves (1.6)}, \quad u \in L^2(0, T). \end{cases} \tag{1.10}$$

In the remaining chapters, we will consistently use the three above examples to illustrate the main theoretical and numerical methods in optimal control of random PDEs.

References

1. Chiba, R.: Stochastic analysis of heat conduction and thermal stresses in solids: a review. In: Kazi, S.N. (ed.) Chapter 9 in Heat Transfer Phenomena and Applications. InTech (2012)
2. de Rocquigny, E., Devictor, N., Tarantola, S.: Uncertainty in Industrial Practice: A Guide to Quantitative Uncertainty Management. John Wiley & Sons, Ltd. (2008)

3. Ghanem, R.G., Spanos, P.D.: Stochastic Finite Elements: A Spectral Approach. Springer-Verlag, New York (1991)
4. Gunzburger, M.D., Webster, C., Zhang, G.: Stochastic finite element methods for partial differential equations with random input data. Acta Numer. **23**, 521–650 (2014)
5. Smith, R.C.: Uncertainty quantification. Theory, implementation, and applications. Comput. Sci. Eng., 12. Society for Industrial and Applied Mathematics (SIAM), Philadelphia, PA (2014)
6. Zhang, D.: Stochastic Methods for Flow in Porous Media: Coping with Uncertainties. Academic Press (2002)

Chapter 2
Mathematical Preliminaires

Everything should be made as simple as possible, but not simpler.

Albert Einstein.

This chapter briefly collects the main theoretical and numerical tools which are needed to the analysis and numerical resolution of control problems under uncertainty. We start with a few basic definitions and general notations. Then, some material on tensor product of Hilbert spaces is reviewed. Finally, the numerical approximation of random fields by using Karhunen-Loève expansions is studied.

2.1 Basic Definitions and Notations

From now on in this text, $D \subset \mathbb{R}^d$, $d = 1, 2, 3$ in applications, is a bounded and Lipschitz domain with boundary ∂D. We denote by $L^p(D)$, $1 \leq p \leq \infty$, the space of all (equivalence classes of) Lebesgue measurable functions $g : D \to \mathbb{R}$ whose norm

$$\|g\|_{L^p(D)} = \begin{cases} \left(\int_D |g(x)|^p \, dx \right)^{1/p}, & p < \infty \\ \text{ess sup}_{x \in D} |g(x)|, & p = \infty \end{cases}$$

is finite. To make explicit the Lebesgue measure and hence avoid confusion, for $1 \leq p < \infty$, $L^p(D)$ will be denoted by $L^p(D; dx)$ in some places in the next section.

Similarly, for $1 \leq p < \infty$,

$$L^p(\partial D) = \left\{ g : \partial D \to \mathbb{R} \text{ measurable} : \int_{\partial D} |g(x)|^p \, ds < \infty \right\},$$

© The Author(s), under exclusive license to Springer Nature Switzerland AG 2018
J. Martínez-Frutos and F. Periago Esparza, *Optimal Control of PDEs under Uncertainty*,
SpringerBriefs in Mathematics, https://doi.org/10.1007/978-3-319-98210-6_2

where ds represents the surface Lebesgue measure on ∂D.

The following Sobolev spaces will be of special importance in the remaining of this book. $H^1(D)$ denotes the space of all functions $y \in L^2(D)$ having first-order partial derivatives (in the sense of distributions) $\partial_{x_j} y$ in $L^2(D)$. This space is endowed with the inner product

$$< y, z >_{H^1(D)} = \int_D yz \, dx + \int_D \nabla y \cdot \nabla z \, dx,$$

which induces the norm

$$\|y\|_{H^1(D)} \left(\int_D (y^2 + |\nabla y|^2) \, dx \right)^{1/2}.$$

As is well-known, $H^1(D)$ is a separable Hilbert space. The space

$$H_0^1(D) = \{ y \in H^1(D) : y \mid_{\partial D} = 0 \},$$

where $y \mid_{\partial D}$ denotes the trace of y, is also considered. In $H_0^1(D)$,

$$\|y\|_{H_0^1(D)} = \left(\int_D |\nabla y|^2 \, dx \right)^{1/2}$$

defines a norm, which is equivalent to the norm in $H^1(D)$.

$(\Omega, \mathcal{F}, \mathbb{P})$ denotes a complete probability space. By complete it is meant that $\mathcal{F}$ contains all $\mathbb{P}$-null sets, i.e., if $B_1 \subset \Omega$ is such that there exists $B_2 \in \mathcal{F}$ satisfying $B_1 \subset B_2$ and $\mathbb{P}(B_2) = 0$, then $B_1 \in \mathcal{F}$. Notice that this concept of completeness means that the measure $\mathbb{P}$ is complete [10, p. 31].

For $1 \le p \le \infty$, $L_{\mathbb{P}}^p(\Omega)$ denotes the space composed of all (equivalence classes of) measurable functions $g : \Omega \to \mathbb{R}$ whose norm

$$\|g\|_{L_{\mathbb{P}}^p(\Omega)} = \begin{cases} \left(\int_\Omega |g(\omega)|^p \, d\mathbb{P}(\omega) \right)^{1/p}, & p < \infty \\ \text{ess sup}_{\omega \in \Omega} |g(\omega)|, & p = \infty \end{cases}$$

is finite. For convenience, $L_{\mathbb{P}}^p(\Omega)$ will also be denoted by $L^p(\Omega; d\mathbb{P}(\omega))$ in the next section.

If $(X, \|\cdot\|_X)$ is a Banach space and $1 \le p \le \infty$, $L_{\mathbb{P}}^p(\Omega; X)$ denotes the Lebesgue-Bochner space composed of all (equivalence classes of) strongly measurable functions $g : \Omega \to X$ whose norm

$$\|g\|_{L^p_{\mathbb{P}}(\Omega;X)} = \begin{cases} \left(\int_\Omega \|g\,(\cdot,\omega)\|_X^p \, d\mathbb{P}\,(\omega)\right)^{1/p}, \, p < \infty \\ \text{ess sup}_{\omega \in \Omega} \|g\,(\cdot,\omega)\|_X, \qquad p = \infty \end{cases}$$

is finite. Since $(\Omega, \mathscr{F}, \mathbb{P})$ is complete, $L^p_{\mathbb{P}}\,(\Omega; X)$ is complete (in the sense that every Cauchy sequence in $L^p_{\mathbb{P}}\,(\Omega; X)$ is a convergent sequence; see [10, pp. 107 and 175]).

Given the probability space $(\Omega, \mathscr{F}, \mathbb{P})$, the probability measure $\mathbb{P}$ induces a distance d in $\mathscr{F}$ as follows: given $E, F \in \mathscr{F}$,

$$d\,(E, F) = \mathbb{P}\,(E\Delta F),$$

where $E\Delta F = (E\,F) \cup (F\,E)$ is the symmetric difference between E and F. The probability space $(\Omega, \mathscr{F}, \mathbb{P})$ is said to be separable if its associated metric space $(\mathscr{F}, d)$ is separable.[1] It is proved that if $(\Omega, \mathscr{F}, \mathbb{P})$ is separable, then $L^p_{\mathbb{P}}\,(\Omega; X)$ is also separable [10, p. 177].

Of particular interest in this book is the case $p = 2$. Precisely, let $(H, < \cdot, \cdot >_H)$ be a separable Hilbert space. Analogously to the scalar case, it may be proved that the vector-valued space $L^2_{\mathbb{P}}\,(\Omega; H)$, with the inner product

$$<f, g>_{L^2_{\mathbb{P}}(\Omega;H)} = \int_\Omega <f(\omega), g(\omega)>_H \, d\mathbb{P}\,(\omega)$$

is also a separable Hilbert space. The induced norm is given by

$$\|f\|_{L^2_{\mathbb{P}}(\Omega;H)} = \left(\int_\Omega \|f(\omega)\|_H^2 \, d\mathbb{P}\,(\omega)\right)^{1/2}.$$

2.2 Tensor Product of Hilbert Spaces

In this section, we review on some definitions and results on tensor product of Hilbert spaces, which shall be very useful in the study of random PDEs. We follow the approach in [17, Chapter 2] and refer the reader to that reference for further details. All the Hilbert spaces considered in this section are assumed to be real and separable.

Definition 2.1 Let $(H_1, < \cdot, \cdot >_{H_1})$, $(H_2, < \cdot, \cdot >_{H_2})$ be two Hilbert spaces. Given $h_1 \in H_1$ and $h_2 \in H_2$, the pure tensor $h_1 \otimes h_2$ is defined as the bilinear form

$$\begin{aligned} h_1 \otimes h_2 : H_1 \times H_2 &\to \mathbb{R} \\ (\varphi_1, \varphi_2) &\mapsto (h_1 \otimes h_2)\,(\varphi_1, \varphi_2) = <h_1, \varphi_1>_{H_1} <h_2, \varphi_2>_{H_2} . \end{aligned} \quad (2.1)$$

Let $\mathscr{H} := \text{span}\,\{h_1 \otimes h_2, \quad h_1 \in H_1, h_2 \in H_2\}$ be the set of all finite linear combinations of such pure tensors. An inner product $< \cdot, \cdot >_{\mathscr{H}}$ is defined on $\mathscr{H}$ by defining it on pure tensors as

---

[1] We recall that a metric space $(\mathscr{F}, d)$ is separable if it has a countable dense subset.

$$\forall \varphi_1 \otimes \varphi_2, \psi_1 \otimes \psi_2, \quad <\varphi_1 \otimes \varphi_2, \psi_1 \otimes \psi_2 >_{\mathscr{H}} = <\varphi_1, \psi_1 >_{H_1} < \varphi_2, \psi_2 >_{H_2},$$

and then by extending it by linearity to $\mathscr{H}$.

It is proved that $< \cdot , \cdot >_{\mathscr{H}}$ is well defined in the sense that $< \varphi, \psi >_{\mathscr{H}}$, with $\varphi, \psi \in \mathscr{H}$, does not depend on which finite combinations are used to express φ and ψ.

Definition 2.2 Let $\left(H_1, < \cdot , \cdot >_{H_1} \right)$, $\left(H_2, < \cdot , \cdot >_{H_2} \right)$ be two Hilbert spaces. The tensor product of H_1 and H_2, denoted by $H_1 \otimes H_2$, is the Hilbert space which is defined as the completion of $\mathscr{H}$ with respect to the inner product $< \cdot , \cdot >_{\mathscr{H}}$.

Proposition 2.1 [17, p. 50] *If* $\{\varphi_i\}_{i \in \mathbb{N}}$ *and* $\left\{\psi_j\right\}_{j \in \mathbb{N}}$ *are orthonormal bases of* H_1 *and* H_2, *respectively, then* $\left\{\varphi_i \otimes \psi_j\right\}_{i,j \in \mathbb{N}}$ *is an orthonormal basis of* $H_1 \otimes H_2$.

Remark 2.1 To simplify notation, both the spaces H_1 and H_2 in the preceding proposition have been taken infinite dimensional. The other cases, i.e. both spaces are finite dimensional or one of them is finite dimensional, are obviously similar.

In this book we are mainly concerned with random fields, i.e., with functions $f(x, \omega) : D \times \Omega \to \mathbb{R}$ which depend on a spatial variable $x \in D$ and on a random event $\omega \in \Omega$. Let us assume that $f(x, \omega) \in L^2(D \times \Omega; dx \times d\mathbb{P}(\omega))$, where dx is the Lebesgue measure on D and $\mathbb{P}$ is the probability measure on Ω. As is usual, $dx \times d\mathbb{P}(\omega)$ denotes the associated product measure. Since $f(x, \omega)$ intrinsically has a different structure with respect to x and ω, thinking on the numerical approximation of such an f, the use of tensor product spaces is very convenient. Indeed, let $\{\varphi_i(x)\}_{i \in \mathbb{N}}$ and $\left\{\psi_j(\omega)\right\}_{j \in \mathbb{N}}$ be orthonormal bases of $L^2(D; dx)$ and $L^2(\Omega; d\mathbb{P}(\omega))$, respectively. By Fubini's theorem, it is not hard to show that $\left\{\varphi_i(x)\psi_j(\omega)\right\}_{i,j \in \mathbb{N}}$ is an orthonormal basis of $L^2(D \times \Omega; dx \times d\mathbb{P}(\omega))$. Now, consider the mapping

$$U : L^2(D; dx) \otimes L^2(\Omega; d\mathbb{P}(\omega)) \to L^2(D \times \Omega; dx \times d\mathbb{P}(\omega)),$$

which is defined on the basis $\left\{\varphi_i(x) \otimes \psi_j(\omega)\right\}_{i,j \in \mathbb{N}}$ by

$$U\left(\varphi_i(x) \otimes \psi_j(\omega)\right) = \varphi_i(x)\psi_j(\omega)$$

and can be uniquely extended to a unitary mapping from $L^2(D; dx) \otimes L^2(\Omega; d\mathbb{P}(\omega))$ onto $L^2(D \times \Omega; dx \times d\mathbb{P}(\omega))$. Hence, if $\varphi(x) = \sum_{i=1}^{\infty} c_i \varphi_i(x) \in L^2(D; dx)$ and $\psi(\omega) = \sum_{j=1}^{\infty} d_j \psi_j(\omega) \in L^2(\Omega; d\mathbb{P}(\omega))$, then one has

$$U(\varphi \otimes \psi) = U\left(\sum_{i,j} c_i d_j \varphi_i(x) \otimes \psi_j(\omega)\right) = \sum_{i,j} c_i d_j \varphi_i(x)\psi_j(\omega) = \varphi(x)\psi(\omega).$$

In this sense, it is said that U defines a *natural* isomorphism between the spaces $L^2(D; dx) \otimes L^2(\Omega; d\mathbb{P}(\omega))$ and $L^2(D \times \Omega; dx \times d\mathbb{P}(\omega))$. We write

$$L^2 (D; dx) \otimes L^2 (\Omega; d\mathbb{P}(\omega)) \cong L^2 (D \times \Omega; dx \times d\mathbb{P}(\omega)). \qquad (2.2)$$

Another interesting use of tensor product spaces is that they enable a nice representation of vector-valued functions. Indeed, let $(H, < \cdot, \cdot >_H)$ be a Hilbert space with basis $\{\varphi_i\}_{i \in \mathbb{N}}$ and consider the Bochner space $L^2_\mathbb{P} (\Omega; H)$. Every function $f \in L^2_\mathbb{P} (\Omega; H)$ may be written as

$$f(\omega) = \lim_{N \to \infty} \sum_{k=1}^N f_k(\omega)\varphi_k, \quad f_k(\omega) = <f(\omega), \varphi_k >_H \quad \text{and } \omega \in \Omega.$$

The mapping

$$\tilde{U} : \sum_{k=1}^N f_k(\omega) \otimes \varphi_k \mapsto \sum_{k=1}^N f_k(\omega)\varphi_k$$

is well-defined from a dense subset of $L^2 (\Omega, d\mathbb{P}(\omega)) \otimes H$ onto a dense subset of $L^2_\mathbb{P} (\Omega; H)$ and it preserves the norms. Hence, $\tilde{U}$ may be uniquely extended to a unitary mapping from $L^2 (\Omega, d\mathbb{P}(\omega)) \otimes H$ onto $L^2_\mathbb{P} (\Omega; H)$. As a consequence, if $f = f(\omega) \in L^2 (\Omega, d\mathbb{P}(\omega))$ and $\varphi \in H$, then $\tilde{U} (f(\omega) \otimes \varphi) = f(\omega)\varphi$. Because of this property, it is said that $\tilde{U}$ is a *natural* isomorphism between $L^2 (\Omega, d\mathbb{P}(\omega)) \otimes H$ and $L^2_\mathbb{P} (\Omega; H)$ and we may write

$$L^2 (\Omega, d\mathbb{P}(\omega)) \otimes H \cong L^2_\mathbb{P} (\Omega; H) \qquad (2.3)$$

Remark 2.2 In the particular case in which $H = L^2 (D; dx)$, by (2.2) and (2.3) we have the isomorphisms

$$L^2_\mathbb{P} (\Omega; L^2 (D)) \cong L^2 (D; dx) \otimes L^2 (\Omega; d\mathbb{P}(\omega)) \cong L^2 (D \times \Omega; dx \times d\mathbb{P}(\omega)).$$

As it will be shown in the next chapter, another very interesting case in which the isomorphism (2.3) applies is when H is a Sobolev space.

All the above discussion is summarized in the following abstract result:

Theorem 2.1 [17, Theorem II.10] *Let (M_1, μ_1) and (M_2, μ_2) be measure spaces so that $L^2 (M_1, d\mu_1)$ and $L^2 (M_2, d\mu_2)$ are separable. Then:*

(a) *There is a unique isomorphism from the tensor space $L^2 (M_1, d\mu_1) \otimes L^2 (M_2, d\mu_2)$ to $L^2 (M_1 \times M_2, d\mu_1 \times d\mu_2)$ so that $\varphi \otimes \psi \mapsto \varphi\psi$.*
(b) *If $(H, < \cdot, \cdot >_H)$ is a separable Hilbert space, then there is a unique isomorphism from $L^2 (M_1, d\mu_1) \otimes H$ to $L^2 (M_1, d\mu_1; H)$ so that $f(x) \otimes \varphi \mapsto f(x)\varphi$.*

Three typical situations to which the abstract results of this section will be applied hereafter are the following:

• Let V be a Sobolev space, e.g. $V = H_0^1 (D)$, and consider the space $L^2_\mathbb{P} (\Omega)$. By part (b) of Theorem 2.1 one has the isomorphism

$$L_{\mathbb{P}}^2 (\Omega; V) \cong L_{\mathbb{P}}^2 (\Omega) \otimes V.$$

- In the context of random PDEs, let us assume that inputs of the PDE depend on a finite number of uncorrelated real-valued random variables $\xi_n : \Omega \to \mathbb{R}$, $1 \le n \le N$. Let us denote by $\Gamma_n = \xi_n (\Omega)$, by $\Gamma = \prod_{n=1}^N \Gamma_n$, and by $\rho : \Gamma \to \mathbb{R}_+$ the joint probability density function (PDF) of the multivariate random variable $\xi = (\xi_1, \ldots, \xi_N)$, which is assumed to factorize as

$$\rho (z) = \prod_{n=1}^N \rho_n (z_n), \quad z = (z_1, \ldots, z_n) \in \Gamma,$$

with $\rho_n : \Gamma_n \to \mathbb{R}_+$. Instead of working in the abstract space $L_{\mathbb{P}}^2 (\Omega)$ one may work in the image space

$$L_\rho^2 (\Gamma) = \left\{ f : \Gamma \to \mathbb{R} : \int_\Gamma f^2 (z) \rho (z) \, dz < \infty \right\},$$

which is equipped with the inner product

$$<f, g >_{L_\rho^2(\Gamma)} = \int_\Gamma f(z) g(z) \rho(z) \, dz.$$

We refer the reader to Sect. 4.1 in Chap. 4 for more details on this passage. Let $\left\{\psi_{p_n} (z_n)\right\}_{p_n=0}^\infty$, $1 \le n \le N$, be an orthonormal basis of $L_{\rho_n}^2 (\Gamma_n)$ composed of a suitable class of orthonormal polynomials. Since $L_\rho^2 (\Gamma) = \bigotimes_{n=1}^N L_{\rho_n}^2 (\Gamma_n)$, a multivariate orthonormal polynomial basis of $L_\rho^2 (\Gamma)$ is constructed as

$$\left\{\psi_p (z) = \prod_{n=1}^N \psi_{p_n} (z_n), \quad z = (z_1, \ldots, z_N) \right\}_{p=(p_1,\ldots,p_N)\in\mathbb{N}^N}.$$

- Let $y = y(z) \in L_\rho^2 \left(\Gamma; H_0^1 (D)\right)$. As it will be illustrated in Chaps. 4 and 5, typically y is the solution of a given elliptic random PDE and one is interested in its numerical approximation. As is usual, the vector valued function $z \mapsto y(\cdot, z) \in H_0^1(D)$ is identified with the real-valued function $D \times \Gamma \ni (x, z) \mapsto y(x, z)$. Having in mind that $L_\rho^2 \left(\Gamma; H_0^1 (D)\right) \cong L_\rho^2 (\Gamma) \otimes H_0^1 (D)$, to approximate numerically $y(x, z)$, a standard finite element space $V_h (D) \subset H_0^1 (D)$ and the space $\mathscr{P}_p (\Gamma) \subset L_\rho^2 (\Gamma)$ composed of polynomials of degree less than or equal to p, are considered. Hence, we look for a numerical approximation of $y \in L_\rho^2 (\Gamma) \otimes H_0^1 (D)$ in the finite dimensional space $\mathscr{P}_p (\Gamma) \otimes V_h (D)$. Let $\left\{\psi_m(z), 0 \le m \le M_p\right\}$ be a basis of $\mathscr{P}_p (\Gamma)$ composed of orthonormal polynomials in $L_\rho^2 (\Gamma)$, and let $\{\phi_n(x), 1 \le n \le N_h\}$ be a basis of shape functions in $V_h (D)$. Then, $\left\{\psi_m \otimes \phi_n, \quad 0 \le m \le M_p, 1 \le n \le N_h\right\}$ is a basis of $\mathscr{P}_p (\Gamma) \otimes V_h (D)$, which is identified with $\{\psi_m(z)\phi_n(x), \quad 0 \le m \le M_p, 1 \le n \le N_h\}$. Thus, an approximation $y_{hp}(x, z) \in \mathscr{P}_p(\Gamma) \otimes V_h(D)$ of $y(x, z)$ is expressed in the form

$$y_{hp}(x, z) = \sum_{n=1}^{N_h} \sum_{m=0}^{p} y_{n,m} \phi_n(x) \psi_m(z).$$

2.3 Numerical Approximation of Random Fields

As the examples in the preceding chapter show, uncertainty in the random inputs of a PDE are typically represented by functions $a : D \times \Omega \to \mathbb{R}$, which depend on a spatial variable $x \in D$ and on a random event $\omega \in \Omega$. Such functions are called *random fields*. Associated to a random field $a(x, \omega)$ it is interesting to consider the following quantities:

- Mean (or expected) function $\overline{a} : D \to \mathbb{R}$, which is defined by

$$\overline{a}(x) = \mathbb{E}[a(x, \cdot)] = \int_{\Omega} a(x, \omega)\, d\mathbb{P}(\omega). \tag{2.4}$$

- Covariance function: $\mathrm{Cov}_a : D \times D \to \mathbb{R}$, which is defined by

$$\mathrm{Cov}_a(x, x') = \mathbb{E}\left[(a(x, \cdot) - \overline{a}(x))(a(x', \cdot) - \overline{a}(x'))\right] \tag{2.5}$$

and expresses the spatial correlation of the random field, i.e., $\mathrm{Cov}_a(x, x')$ is the covariance of the random variables $\omega \mapsto a(x, \omega)$ and $\omega \mapsto a(x', \omega)$.
- Variance function: $\mathrm{Var}_a(x) = \mathrm{Cov}_a(x, x)$.

The random field a is said to be *second order* if $\mathrm{Var}_a(x) < \infty$ for all $x \in D$. A particular type of random fields, which are widely used to model uncertainty in systems, is the class of Gaussian fields. A random field $a(x, \omega)$ is said to be Gaussian if for any $M \in \mathbb{N}$ and $x^1, x^2, \ldots, x^M \in D$ the $\mathbb{R}^M$-valued random variable $X(\omega) = (a(x^1, \omega), a(x^2, \omega), \ldots, a(x^M, \omega))$ follows the multivariate Gaussian distribution $\mathscr{N}(\mu, C)$, where

$$\mu = (\overline{a}(x^1), \ldots, \overline{a}(x^M)) \quad \text{and} \quad C_{ij} = \mathrm{Cov}_a(x^i, x^j), \quad 1 \le i, j \le M.$$

For computational purposes, it is very convenient to express a random field $a(x, \omega)$ by means of a finite number of random variables $\{\xi_n(\omega)\}_{1 \le n \le N}$, with $N \in \mathbb{N}_+ = \{1, 2, 3, \ldots\}$. Among the different possibilities, Karhunen-Loève expansion is a very usual choice to represent a random field.

2.3.1 Karhunen-Loève Expansion of a Random Field

As it will be illustrated along this text, it is very appealing to have a representation of a random field in the form

$$a(x, \omega) = \bar{a}(x) + \sum_{n=1}^{\infty} b_n(x) \hat{\xi}_n(\omega), \qquad (2.6)$$

with *pairwise uncorrelated*[2] random variables $\left\{ \hat{\xi}_n(\omega) \right\}_{n=1}^{\infty}$ having, for convenience, zero mean and σ_n^2 variances, and $L^2(D)$-orthonormal functions $\{b_n(x)\}_{n=1}^{\infty}$.

To derive how the spatial functions $b_n(x)$ and the random variables $\hat{\xi}_n(\omega)$ look like, we continue our discussion in this section proceeding in a formal way. If (2.6) holds, then the covariance function (2.5) admits (at least formally) the representation

$$\begin{aligned}
\text{Cov}_a(x, x') &= \int_{\Omega} \sum_{i=1}^{\infty} b_i(x) \hat{\xi}_i(\omega) \sum_{n=1}^{\infty} b_n(x') \hat{\xi}_n(\omega) \, d\mathbb{P}(\omega) \\
&= \sum_{i=1}^{\infty} \sum_{n=1}^{\infty} b_i(x) b_n(x') \int_{\Omega} \hat{\xi}_i(\omega) \hat{\xi}_n(\omega) \, d\mathbb{P}(\omega) \\
&= \sum_{n=1}^{\infty} \sigma_n^2 b_n(x) b_n(x').
\end{aligned}$$

By multiplying this expression by $b_j(x')$ and integrating in D, a direct (still formal) computation leads to

$$\int_D \text{Cov}_a(x, x') b_j(x') \, dx' = \sigma_j^2 b_j(x),$$

which means that $\left\{ \sigma_n^2, b_n(x) \right\}_{n=1}^{\infty}$ are the eigenpairs associated to the operator

$$\psi \mapsto \int_D \text{Cov}_a(x, x') \psi(x') \, dx', \quad \psi \in L^2(D).$$

Denoting by $\xi_n = \frac{1}{\sigma_n} \hat{\xi}_n$, $1 \leq n \leq \infty$, the random variables ξ_n have zero mean and unit variance. Then, (2.6) rewrites as

$$a(x, \omega) = \bar{a}(x) + \sum_{n=1}^{\infty} \sigma_n b_n(x) \xi_n(\omega). \qquad (2.7)$$

---

[2] We recall that a family of random variables $\left\{ \hat{\xi}_n(\omega) \right\}_{n=1}^{\infty}$ having zero mean is said to be pairwise uncorrelated if $\int_{\Omega} \hat{\xi}_i(\omega) \hat{\xi}_j(\omega) \, d\mathbb{P}(\omega) = 0$ for $i \neq j$.

Hence,

$$\int_D [a(x,\omega) - \bar{a}(x)] b_n(x)\, dx = \sum_{i=1}^{\infty} \sigma_i \xi_i(\omega) \int_D b_i(x) b_n(x)\, dx = \sigma_n \xi_n(\omega),$$

that is,

$$\xi_n(\omega) = \frac{1}{\sigma_n} \int_D [a(x,\omega) - \bar{a}(x)] b_n(x)\, dx.$$

These formal computations can be made rigorous, as the following results state. Before that, let us recall that a function $C : D \times D \to \mathbb{R}$ is symmetric if $C(x,x') = C(x',x)$ for all $x, x' \in D$ and C is non-negative definite if for any $x_j \in D$ and $\alpha_j \in \mathbb{R}$, $1 \leq j \leq N$, the inequality

$$\sum_{j,k=1}^{N} \alpha_j \alpha_k C(x_j, x_k) \geq 0$$

holds.

Theorem 2.2 (Mercer) [14, Theorem 1.8] *Let $D \subset \mathbb{R}^d$ be a bounded domain and let $C : \overline{D} \times \overline{D} \to \mathbb{R}$ be a continuous, symmetric and non-negative definite function. Consider the integral operator $\mathscr{C} : L^2(D) \to L^2(D)$ defined by*

$$(\mathscr{C}\psi)(x) = \int_D C(x,x') \psi(x')\, dx'. \tag{2.8}$$

Then, there exist a sequence of, continuous in $\overline{D}$, eigenfunctions $\{b_n\}_{n=1}^{\infty}$ of $\mathscr{C}$, which can be chosen orthonormal in $L^2(D)$, such that the corresponding sequence of eigenvalues $\{\lambda_n\}_{n=1}^{\infty}$ is positive. Moreover,

$$C(x,x') = \sum_{n=1}^{\infty} \lambda_n b_n(x) b_n(x'), \quad x, x' \in \overline{D}, \tag{2.9}$$

where the series converges absolutely and uniformly on $\overline{D} \times \overline{D}$.

Theorem 2.3 (Karhunen-Loève expansion) [14, Theorems 5.28, 5.29, 7.52, 7.53] *Let $D \subset \mathbb{R}^d$ be a bounded domain and let $a = a(x,\omega)$ be a random field with continuous covariance[3] function $Cov_a : \overline{D} \times \overline{D} \to \mathbb{R}$. Then, $a(x,\omega)$ admits the Karhunen-Loève expansion*

---

[3] We recall that not any function may be taken as the covariance function of a random field. To be a valid covariance function it must be, in particular, non-negative definite. See [14, Theorem 5.18] for more details on this issue. All covariance functions considered in this text are assumed to be continuous, symmetric and non-negative definite.

$$a\left(x, \omega\right) = \bar{a}\left(x\right) + \sum_{n=1}^{\infty} \sqrt{\lambda_n} b_n\left(x\right) \xi_n\left(\omega\right), \tag{2.10}$$

where the sum converges in $L_{\mathbb{P}}^2\left(\Omega; L^2\left(D\right)\right)$. $\{\lambda_n, b_n\left(x\right)\}_{n=1}^{\infty}$ are the eigenpairs of the operator $\mathscr{C}$ defined in Mercer's theorem, with $C\left(x, x'\right) = Cov_a\left(x, x'\right)$, and

$$\xi_n\left(\omega\right) = \frac{1}{\sqrt{\lambda_n}} \int_D \left[a\left(x, \omega\right) - \bar{a}\left(x\right)\right] b_n\left(x\right) dx \tag{2.11}$$

are pairwise uncorrelated random variables with zero mean and unit variance. If the random field $a(x, \omega)$ is Gaussian, then $\xi_n \sim \mathscr{N}\left(0, 1\right)$ are independent and identically distributed (i.i.d.) standard Gaussian variables.

Remark 2.3 Note that, thanks to Theorem 2.3, a Gaussian field $a(x, \omega)$ is completely determined by its expected function $\bar{a}\left(x\right)$ and its covariance function. Indeed, from $Cov_a\left(x, x'\right)$ one computes the eigenpairs $\{\lambda_n, b_n(x)\}_{n=1}^{\infty}$ of the operator (2.8). Thus, if $\bar{a}\left(x\right)$ is known, then the KL expansion (2.10) determines the Gaussian field $a(x, \omega)$ since the random variables that appear in (2.10) are i.i.d. standard Gaussian variables. As a consequence, if the mean and covariance functions of the Gaussian field $a(x, \omega)$ are properly estimated from experimental data, then a is completely characterized (see also Remark 2.4 for more details on this issue).

For numerical simulation purposes, the expansion (2.10) is always truncated at some large enough term N, i.e., the random field $a\left(x, \omega\right)$ is approximated as

$$a\left(x, \omega\right) \approx a_N\left(x, \omega\right) = \bar{a}\left(x\right) + \sum_{n=1}^{N} \sqrt{\lambda_n} b_n\left(x\right) \xi_n\left(\omega\right). \tag{2.12}$$

The convergence rate of $a_N\left(x, \omega\right)$ depends on the decay of the eigenvalues λ_n, which in turn depends on the smoothness of it associated covariance function. Precisely, for piecewise Sobolev regular covariance functions, the eigenvalue decay is algebraic. For piecewise analytic covariance functions, the rate decay is quasi-exponential (see [19] and the references therein).

A measure of the relative error committed by truncation of a KL expansion is given by the so-called *variance error* [14, 18]

$$E_N\left(x\right) = \frac{Var_{a-a_N}\left(x\right)}{Var_a\left(x\right)} = \frac{C\left(x, x\right) - \sum_{n=1}^{N} \lambda_n b_n^2\left(x\right)}{C\left(x, x\right)} = 1 - \frac{\sum_{n=1}^{N} \lambda_n b_n^2\left(x\right)}{C\left(x, x\right)}, \tag{2.13}$$

where, to alleviate the notation, from now on $C = Cov_a$ is the covariance function of a. Second equality in (2.13) is easily obtained from Theorems 2.2 and 2.3. See [14, Corollary 7.54] for more details. A global measure of this relative error, which is called *mean error variance*, is

$$\overline{E}_N = \frac{\int_D \text{Var}_{a-a_N}(x)\, dx}{\int_D \text{Var}_a(x)\, dx} = \frac{\int_D C(x,x)\, dx - \sum_{n=1}^N \lambda_n}{\int_D C(x,x)\, dx} = 1 - \frac{\sum_{n=1}^N \lambda_n}{\int_D C(x,x)\, dx}.$$

$$(2.14)$$

For some specific covariance functions and simple geometries of the physical domain D, it is possible to find analytical expressions of KL expansions, as the following example shows.

Example 2.1 (Exponential Covariance) Consider a Gaussian field $a = a(x, \omega)$ with zero mean, unit variance and having the following exponential covariance function:

$$C(x, x') = e^{-|x-x'|/L_c}, \quad x, x' \in D = (0, L),$$

$$(2.15)$$

where L_c is the so-called correlation length. The eigenpairs $\{\lambda_n, b_n(x)\}_{n=1}^\infty$ associated to $C(x, x')$ are obtained by solving the spectral problem

$$\int_0^x e^{-\frac{x-x'}{L_c}} b(x')\, dx' + \int_x^L e^{-\frac{x'-x}{L_c}} b(x')\, dx' = \lambda b(x).$$

$$(2.16)$$

By differentiating twice and by evaluating $b'(x)$ at the end points $x = 0, L$, the spectral problem (2.16) takes the form

$$\begin{cases} b''(x) + \frac{1}{L_c}\left(\frac{2}{\lambda} - \frac{1}{L_c}\right) b(x) = 0, & 0 < x < L \\ b'(0) = \frac{1}{L_c} b(0), \quad b'(L) = -\frac{1}{L_c} b(L). \end{cases}$$

$$(2.17)$$

Denoting by $\ell = L/L_c$, the eigenvalues of (2.17) are given by

$$\lambda_n = \frac{2\ell^2 L_c}{\ell^2 + \gamma_n^2},$$

$$(2.18)$$

where γ_n, $n \in \mathbb{N}$, are the positive solutions of the transcendental equation

$$\cot(\gamma_n) = \frac{\gamma_n^2 - \ell^2}{2\ell\gamma_n}.$$

$$(2.19)$$

Its associated normalized eigenfunctions are

$$b_n(x) = \sqrt{\frac{2}{\ell^2 + \gamma_n^2 + 2\ell}} \left[\gamma_n \cos\left(\frac{\gamma_n x}{L}\right) + \ell \sin\left(\frac{\gamma_n x}{L}\right) \right], \quad 0 < x < L. \quad (2.20)$$

The energy of a is given by

$$\|a\|_{L^2(D)\otimes L^2_{\mathbb{P}}(\Omega)}^2 = \int_0^L \text{Var}_a(x)\, dx = \int_0^L C(x,x)\, dx = L.$$

$$(2.21)$$

Consider the N-term truncation $a_N(x, \omega) = \sum_{n=1}^{N} \sqrt{\lambda_n} b_n(x) \xi_n(\omega)$, where the eigenvalues are ordered in decreasing order, i.e., $\lambda_1 \geq \lambda_2 \geq \cdots \geq \lambda_N$. Taking into account that $\{b_n(x)\}_{n=1}^{\infty}$ are orthonormal in $L^2(D)$ and that $\xi_n \sim \mathcal{N}(0, 1)$ are independent, a direct computation shows that

$$\|a_N\|^2_{L^2(D) \otimes L^2_{\mathbb{P}}(\Omega)} = \sum_{n=1}^{N} \lambda_n. \tag{2.22}$$

Hence, if one aims to capture a certain amount, e.g. 90%, of the energy of the field a, then the truncation term N must be chosen as to satisfy the condition $\overline{E}_N \leq 0.1$, with $\overline{E}_N$ given by (2.14), which, in this example, corresponds to

$$\sum_{n=1}^{N} \lambda_n \geq 0.9L.$$

Figure 2.1a shows the first 100 eigenvalues obtained using (2.18), with $L = 1$ and different values of L_c. The transcendental equation (2.19) is solved by the classical Newton method. Figure 2.1b displays the first 5 eigenfunctions for $L_c = 0.5$.

The rate decay of eigenvalues determines how fast the variance error and the mean error variance are reduced. This fact is observed in Fig. 2.2, where these errors are depicted as a function of N and for several values of the correlation length L_c. One can observe how the number of terms needed to reduce the relative error to a desired threshold increases as $L_c \to 0$. This issue is of paramount importance from a computational point of view since the number of terms of the KL expansion and thus the number of simulations required for an accurate representation of the field

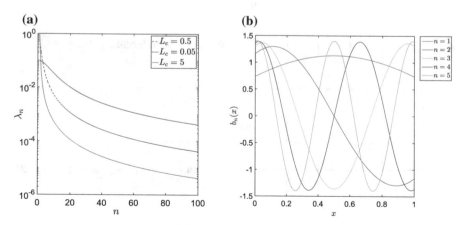

Fig. 2.1 Example 2.1: (**a**) Eigenvalues λ_n, $1 \leq n \leq 100$ for the correlation lengths $L_c = 0.05$ (dash red line), $L_c = 0.5$ (dot-dash blue line) and $L_c = 5$ (solid green line). (**b**) Eigenfunctions for $1 \leq n \leq 5$ and $L_c = 0.5$

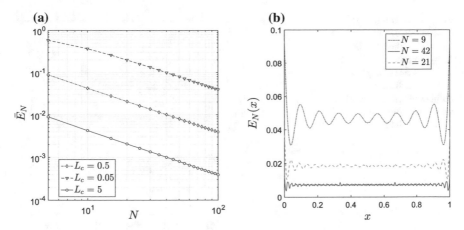

Fig. 2.2 Example 2.1: (**a**) Mean error variance $\overline{E}_N$ for $1 \leq N \leq 100$ and $L_c = 0.05$ (dash line), $L_c = 0.5$ (dot-dash line) and $L_c = 5$ (solid line), and (**b**) variance error $E_N(x)$ for $L_c = 0.5$ and $N = 9, 21, 42$

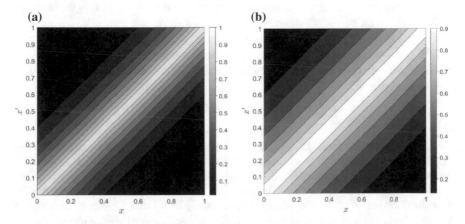

Fig. 2.3 Example 2.1: (**a**) Covariance of $a(x, \omega)$, and (**b**) Covariance of $a_{42}(x, \omega)$

would be large in case the spectrum decays slowly. Notice also that the largest values of $E_N(x)$ in Fig. 2.2b are located at the end points of the interval. This is due to the fact that since the eigenvalues are ordered in decreasing order, the dominant term in $E_N(x)$ is the first one, whose associated eigenfunction $b_1(x)$ takes its minimum values at $x = 0, 1$ (see Fig. 2.1b).

In the same vein, Fig. 2.3 displays the covariance functions of the random field $a(x, \omega)$ and of its truncation at the 42-th term.

Three realizations of $a_N(x, \omega)$ are shown in Fig. 2.4. Observe that, for small values of the spatial correlation length, the random field tends to be rough. When $L_c \to 0$ the random field can be viewed as a infinite set of uncorrelated random variables.

Fig. 2.4 Example 2.1: realizations of $a_N(x, \omega)$ for $L_c = 0.05$ (dash red line), $L_c = 0.5$ (dot-dash blue line) and $L_c = 5$ (solid green line). In each case, the number of terms N has been fixed to capture the 95% of the energy field

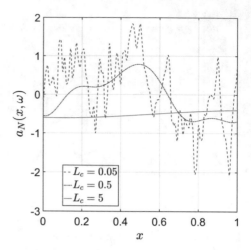

Conversely, the random field becomes smoother as the correlation length increases. In the limit, when $L_c \to \infty$, the random field is uniform.

For more complicated geometries and/or covariance functions, the eigenpairs $\{\lambda_n, b_n(x)\}$ need to be approximated numerically because in these cases it is not possible to have explicit solutions of the spectral problem: find $\lambda \in \mathbb{R}$ and $b \in L^2(D)$, $b \neq 0$, such that

$$\int_D C(x, x') b(x') \, dx' = \lambda b(x), \quad x \in D. \tag{2.23}$$

Several numerical methods have been proposed in the literature [3]. Since for the class of problems considered in this book, a mesh of the physical domain D is assumed to be readily available, here we describe a discetized version of (2.23) based on the finite element method (FEM). Let $V_h(D) \subset L^2(D)$ be a finite element space associated to a mesh of size h of D and let $\{\varphi_1, \varphi_2, \ldots, \varphi_M\}$ be a basis of shape functions of $V_h(D)$. Then, we look for approximated eigenvalues $\hat{\lambda}_n$ and eigenfunctions $\hat{b}_n(x) = \sum_{k=1}^{M} b_n^k \varphi_k(x)$ by solving the generalized matrix eigenvalue problem

$$\sum_{k=1}^{M} b_n^k \int_D \int_D C(x, x') \varphi_k(x') \varphi_i(x) \, dx' dx = \hat{\lambda}_n \sum_{k=1}^{M} b_n^k \int_D \varphi_k(x) \varphi_i(x) \, dx, \quad 1 \leq i \leq M, \tag{2.24}$$

which, in matrix form, is written as

$$\mathbf{C}\mathbf{b}_n = \hat{\lambda}_n \mathbf{B}\mathbf{b}_n, \tag{2.25}$$

with $\mathbf{b}_n = \left[b_n^1, \ldots, b_n^M \right]^T$,

Fig. 2.5 Example 2.2.
Relative error of eigenvalues
for $0 \leq n \leq 42$ and different
sizes of the finite element
mesh. $L_c = 0.5$ in (2.15)

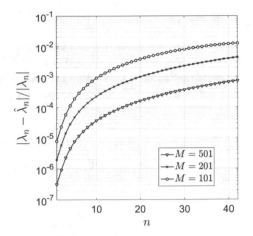

$$[\mathbf{C}]_{ij} = \int_D \int_D C\left(x, x'\right) \varphi_i\left(x'\right) \varphi_j\left(x\right) dx'dx, \quad \text{and} \quad [\mathbf{B}]_{ij} = \int_D \varphi_i\left(x\right) \varphi_j\left(x\right) dx.$$

For error estimates of this numerical method we refer to [4, 19]. The variance error associated to the finite element discretization is given by (2.13), with λ_n replaced by $\hat{\lambda}_n$, and $b_n\left(x\right)$ by $\hat{b}_n\left(x\right)$. The same applies for (2.14). Let us now revisit Example 2.1.

Example 2.2 (Continuation of Example 2.1) For the covariance function (2.15), the generalized matrix eigenvalue problem (2.25) is solved by using Lagrange $P1$ finite elements on a uniform mesh of size $h = 1/(M - 1)$. Figure 2.5 shows the relative error, w.r.t. the eigenvalues λ_n computed in Example 2.1, committed for $M = 101, 201, 501$.

Remark 2.4 (**Log-normal random fields**) When random fields or random variables are used to model input data of a mathematical model which are, in nature, *positive*, Gaussian fields may not be the best choice because negative values have positive probability. In addition, it is observed [13] that experimental data for most of these parameters are log-normal distributed, which means that the *logarithm* (in practice, $\log_{10}$ although in theory the natural logarithmic log is used) of the original data are normally (or Gaussian) distributed. The reason for this behaviour is that many physical laws involve products of physical magnitudes. Thus, the *additive* hypothesis of small errors, which state that if random variation is the sum of many small random effects, then a normal distribution is the result, does not apply. The *multiplicative* version of the Central Limit Theorem ("Geometric means of (non-log-normal) variables are approximately log-normally distributed") must be used instead. We refer the reader to [1, 13] for more details on this passage. This supports the *multiplicative* hypothesis of small errors: if random variation is the product of several random effects, then a log-normal distribution must be the result. A log-normal random field is expressed in the form

$$a(x, \omega) = e^{\mu(x) + \sigma(x)z(x,\omega)}, \tag{2.26}$$

where $z(x, \omega)$ is a standard Gaussian random field with zero mean and unit variance, the so-called geometric mean $\mu_a^*(x) = e^{\mu(x)}$ is a *scale parameter*, and the geometric standard deviation $\sigma_a^*(x) = e^{\sigma(x)}$ is a *shape parameter*. The mean $\bar{a}(x)$ and variance $\text{Var}_a(x)$ functions of the field a are given by

$$\bar{a}(x) = e^{\mu(x) + \frac{1}{2}\sigma^2(x)}, \quad \text{Var}_a(x) = e^{2\mu(x) + \sigma^2(x)}\left(e^{\sigma^2(x)} - 1\right).$$

Hence, if the mean $\bar{a}(x)$ and variance $\text{Var}_a(x)$ functions of the field a are known, then $\mu(x)$ and $\sigma(x)$ are obtained through the transformation

$$\mu(x) = \log\left(\frac{\bar{a}^2(x)}{\sqrt{\bar{a}^2(x) + \text{Var}_a(x)}}\right), \quad \sigma^2(x) = \log\left(1 + \frac{\text{Var}_a(x)}{\bar{a}^2(x)}\right). \tag{2.27}$$

In practice, $\mu_a^*(x) = e^{\mu(x)}$ and $\sigma_a^* = e^{\sigma(x)}$ are more useful than the usual arithmetic mean and arithmetic standard deviation of $a(x, \omega)$. For instance, for a log-normal random variable, the interval $\left[\mu_a^*/\left(\sigma_a^*\right)^2, \mu_a^*\left(\sigma_a^*\right)^2\right]$ covers a probability of 95.5%.

Due to its relevance in the problems addressed along the book, the discretization process of a two-dimensional (2D), in space, log-normal random field with squared exponential covariance is presented in the following example.

Example 2.3 (2D Log-normal random field) Let $z(x, \omega)$ be a Gaussian random field with zero mean, unit variance and the following isotropic squared exponential covariance function:

$$C(x, x') = exp\left[-\sum_{i=1}^{2}\frac{(x_i - x_i')^2}{L_i^2}\right], \quad x = (x_1, x_2), x' = (x_1', x_2') \in D = (0, 1)^2, \tag{2.28}$$

where L_1 and L_2 are the correlation lengths in the two spatial directions. Eigenvalues and eigenfunctions associated to C may be numerically approximated by using the FEM described above. However, since the covariance function (2.28) is expressed as

$$C(x, x') = \prod_{j=1}^{2} C_j\left(x_j, x_j'\right),$$

with $C_j\left(x_j, x_j'\right) = exp\left[-\left(x_j - x_j'\right)^2/L_j^2\right]$, $x_j, x_j' \in (0, 1)$, $j = 1, 2$, its eigenvalues (and eigenfunctions) are expressed as the product of the eigenvalues (eigenfunctions) associated to the one-dimensional (1D) covariance functions $C_j\left(x_j, x_j'\right), j = 1, 2$. See [14, Example 7.56] for a similar situation. From a truncated KL expansion of the Gaussian field $z(x, \omega)$ one obtains an approximation of $a(x, \omega)$ in the form

$$a\left(x, \omega\right) \approx a_N\left(x, \omega\right) = e^{\mu(x) + \sigma(x) \sum_{n=1}^{N} \sqrt{\lambda_n} b_n(x) \xi_n(\omega)}. \tag{2.29}$$

Even if the random field (2.29) takes positive values, it is not completely satis-factory from a mathematical point of view because since the random variables $\xi_n(\omega)$ are Gaussian, $a_N(x, \omega)$ may approximate to zero and to ∞. Hence, when $a_N(x, \omega)$ appears in the principal part of an elliptic differential operator, the uni-form (w.r.t. ω) ellipticity condition, which is required in most of PDEs, does not hold (see next chapter). A natural way of overcoming this technical difficulty is by truncating the Gaussians. Indeed, instead of using the probability density function $\hat{\phi}(s) = \frac{1}{\sqrt{2\pi}} e^{-s^2/2}$, $s \in \mathbb{R}$, of the Gaussian variable $\xi_n(\omega)$, one considers, for some large enough $d > 0$, the probability density function

$$\phi\left(s\right) = \begin{cases} \frac{\hat{\phi}(s)}{\Phi(d) - \Phi(-d)}, & -d \leq s \leq d \\ 0, & \text{otherwise,} \end{cases} \tag{2.30}$$

where $\Phi\left(s\right) = \int_{-\infty}^{s} \hat{\phi}\left(t\right) dt$, $s \in \mathbb{R}$, is the cumulative density function of the normal distribution. For instance, since ξ_n have unit variance, for $d = 2$ a probability of 0.9545 is covered.

The Gaussian random field $z(x, \omega)$ is discretized using a Karhunen-Loève expansion with 15 terms. Figure 2.6 shows the first four eigenfunctions. This random field discretization is able to capture the 99.7% of the total energy of the field providing a good approximation of the variance of the field, which is depicted in Fig. 2.7.

Figure 2.8 shows some realizations of $z_N(x, \omega)$ and $a_N(x, \omega)$, with $N = 15$, for Gaussian and log-normal isotropic ($L_1 = L_2 = 0.5$) and anisotropic ($L_1 = 0.05$, $L_2 = 0.5$) random fields. The log-normal random field is assumed to have mean and variance functions both equal to 1. Hence, from (2.27) it is obtained $\mu = -0.3466$ and $\sigma = 0.8326$. Figure 2.8c, d show how the directional preference of the anisotropy is evident. The realizations appears streched from left to right due to the anisotropy of the covariance function.

2.4 Notes and Related Software

In Sect. 2.1 we have introduced some basic definitions and properties on Functional Analysis and Probability Theory. The reader is referred to [15] for details concerning Sobolev spaces and to [7, 10, 16] for definitions, properties and proofs, concerning probability spaces.

Apart from KL expansion, several methods of discretization of random fields have been proposed in the literature. These methods can be broadly classified into three main groups [18]. The first group consists of methods that discretize the random field by a finite set of random variables averaged over local domains, such as the *weighted integral method* [6] and the *local averaging method* [20]. The second group is based on point discretization methods in which the random variables represent

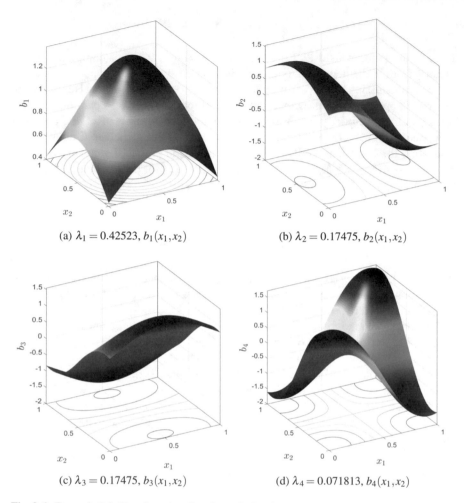

(a) $\lambda_1 = 0.42523$, $b_1(x_1,x_2)$ (b) $\lambda_2 = 0.17475$, $b_2(x_1,x_2)$

(c) $\lambda_3 = 0.17475$, $b_3(x_1,x_2)$ (d) $\lambda_4 = 0.071813$, $b_4(x_1,x_2)$

Fig. 2.6 Example 2.3. First four eigenfunctions of $z(x, \omega)$

the values of the random field at some points. The *midpoint method* [11] or the *optimal linear estimation method (OLE)* [12] are some of the methods included in the category. The third group is composed of series expansion methods which approximate the random field by a finite sum of products of deterministic spatial functions and random variables. The *polynomial chaos (PC) expansion* [21] and the above mentioned *Karhunen-Loève expansion* [14] belong to this category.

The KL expansion has received much attention in the literature because it is optimal in the sense that it minimizes the mean square truncation error [9, Chap. 2].

Methods for numerically approximate eigenvalues and eigenfunctions associated to covariance functions may be broadly classified into three categories [3]: *Nyström methods* [2], *Degenerate kernel methods* and *Projection methods*. An in-depth

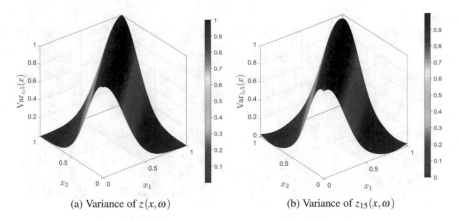

(a) Variance of $z(x, \omega)$ (b) Variance of $z_{15}(x, \omega)$

Fig. 2.7 Example 2.3: (**a**) Variance of of $z(x, \omega)$, and (**b**) variance of $z_{15}(x, \omega)$

discussion of all these methods lies beyond the scope of this book. The reader is referred to [3, 18] for a comprehensive overview.

The exponential and squared exponential covariance functions introduced in this chapter are widely used in applications (see, e.g. [5, Chap. 4] and [8]). The exponential covariance function generates a random field which is only mean-square continuous. The random field constructed from the squared exponential covariance function is mean-square differentiable. We refer to [14, Sect. 5.5] for more details on regularity issues of random fields.

To illustrate the implementation issues of the examples presented along the chapter, several MATLAB scripts are provided accompanying this book. For the interested reader, below are some references to open source code for UQ using random fields:

- FERUM - Finite Element Reliability Using MATLAB.
  http://www.ce.berkeley.edu/projects/ferum/.
- OpenTURNS - Free Software for uncertainty quantification, uncertainty propagation, sensitivity analysis and meta-modelling (written in C++/Python).
  http://www.openturns.org/.
- UQlab - Computational tools that provide efficient, flexible and easy to use programs for UQ on a variety of applications (written in MATLAB).
  http://www.uqlab.com/.
- Uribe, Felipe (2015). MATLAB software for stochastic finite element analysis.
  http://www.mathworks.com/matlabcentral/profile/authors/2912338-felipe-uribe.

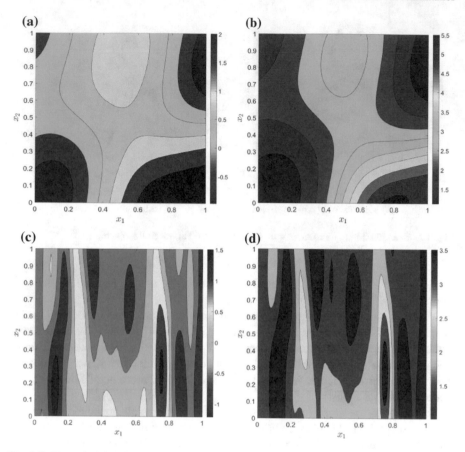

Fig. 2.8 Example 2.3: left panel shows realizations of $z_{15}(x, \omega)$ with correlation lengths $[L_1, L_2] = [0.5, 0.5]$ **(a)** and $[L_1, L_2] = [0.05, 0.5]$ **(c)**. Right panel shows realizations of $a_{15}(x, \omega)$ with correlation lengths $[L_1, L_2] = [0.5, 0.5]$ **(b)** and $[L_1, L_2] = [0.05, 0.5]$ **(d)**

References

1. Aitchison, J., Brown, J. A. C.: The Lognormal Distribution, with Special Reference to its Uses in Economics. Cambridge University Press (1957)
2. Atkinson, K.: The Numerical Solution of Integral Equations of the Second Kind. Cambridge University Press (1997)
3. Betz, W., Papaioannou, I., Straub, D.: Numerical methods for the discretization of random fields by means of the Karhunen-Loève expansion. Comput. Methods Appl. Mech. Engrg. **271**, 109–129 (2014)
4. Chatelin, F.: Spectral Approximation of Linear Operators. Academic Press (1983)
5. Cressie, N., Wikle, C.K.: Statistics for Spatio-Temporal Data. Wiley Series in Probability and Statistics. John Wiley & Sons Inc., Hoboken, NJ (2011)
6. Deodatis, G., Shinozuka, M.: Weighted integral method. II: Response variability and reliability. J. Eng. Mech. **117**(8), 1865–1877 (1991)

7. Durrett, R.: Probability: Theory and Examples. 4 edn. Cambridge Series in Statistical and Probabilistic Mathematics, vol. 31. Cambridge University Press, Cambridge (2010)
8. Gneiting, T., Guttorp, P.: Continuous Parameter Stochastic Process Theory. Handbook of Spatial Statistics, 17–28, Chapman & Hall/CRC Handb. Mod. Stat. Methods, CRC Press, Boca Raton, FL (2010)
9. Ghanem, R.G., Spanos, P.D.: Stochastic Finite Elements: A Spectral Approach. Springer, New York (1991)
10. Halmos, P.: Measure Theory. Springer, New York (1950)
11. Kiureghian, A.D., Ke, J.-B.: The stochastic finite element method in structural reliability. Probab. Eng. Mech. **3**(2), 83–91 (1988)
12. Li, C.-C., Kiureghian, A.D.: Optimal discretization of random fields. J. Eng. Mech. ASCE. **119**(6), 1136–1154 (1993)
13. Limpert, E., Stahel, W. A., Abbt, M.: The log-normal distribution across sciences: keys and clues. BioScience **51**(5) (2001)
14. Lord, G.L., Powell, C.E., Shardlow, T.: An Introduction to Computational Stochastic PDEs. Cambridge University Press (2014)
15. Nečas, J.: Les Méthodes Directes en Théorie des Equations Elliptiques. Éditeurs Academia, Prague (1967)
16. Rao, M.M., Swift, R.J.: Probability Theory with Applications. Vol. 582. 2 edn. Springer (2006)
17. Reed, M., Simon, B.: Methods of Modern Mathematical Physics, vol. 1. Academic Press, Functional Analysis (1980)
18. Sudret, B., Kiureghian, A. D.: Stochastic Finite Element Methods and Reliability: A State-of-the-Art Report. Technical Report UCB/SEMM-2000/08, Department of Civil and Environmental Engineering, University of California, Berkeley (2000)
19. Todor, R.A.: Robust eigenvalue computation for smoothing operators. SIAM J. Numer. Anal. **44**(2), 865–878 (2006)
20. Vanmarcke, E., Grigoriu, M.: Stochastic finite element analysis of simple beams. J. Eng. Mech. **109**(5), 1203–1214 (1983)
21. Wiener, N.: The homogeneous chaos. Amer. J. Math. **60**(4), 897–936 (1938)

Chapter 3
Mathematical Analysis of Optimal Control Problems Under Uncertainty

> ... the methods of mathematical analysis have allowed to
> rigorously identify the missing elements for the models to be
> complete, facilitating the reliable use of computational methods.
>
> Jacques Louis Lions.
> Translated from Spanish. Bolletín of SEMA 15, 2000.

This chapter is focussed on existence theory for the solutions of robust and risk averse optimal control problems. As a first step, the classical variational theory of deterministic PDEs is extended to the case of random PDEs. This variational theory may be developed by using either the formalism of tensor product of Hilbert spaces or abstract functions, i.e., functions with values in Banach or Hilbert spaces. However, tensor products of Hilbert spaces have the advantage that the numerical approximation of such random PDEs becomes very natural in such a formalism.

3.1 Variational Formulation of Random PDEs

Well-posedness of Random Partial Differential Equations (RPDEs) may be studied by working only on the spatial domain D and thus, by considering a deterministic PDE for each realization of the random event $\omega \in \Omega$. Another different approach consists of working in $D \times \Omega$. Hence, solutions of RPDEs are looked for in a Hilbert space such as $L_{\mathbb{P}}^2 (\Omega; V)$, with V a suitable Sobolev space of functions defined in D. The latter approach has the advantage that solutions of random PDEs are found in $L_{\mathbb{P}}^2 (\Omega; V)$ so that their mean and variance are well-defined. For this reason, a formulation in $D \times \Omega$ is presented next.

Since the goal of this section is to illustrate how to deal with the new ingredient of working with random input parameters, our study is limited to the three toy models

© The Author(s), under exclusive license to Springer Nature Switzerland AG 2018
J. Martínez-Frutos and F. Periago Esparza, *Optimal Control of PDEs under Uncertainty*,
SpringerBriefs in Mathematics, https://doi.org/10.1007/978-3-319-98210-6_3

presented in Chap. 1, which, however, cover the usual cases of elliptic equations, and first and second order in time PDEs. Other linear RPDEs may be studied in a similar way.

3.1.1   The Laplace-Poisson Equation Revisited I

Consider the elliptic problem

$$
\begin{cases}
-\text{div}\,(a(x,\omega)\nabla y(x,\omega)) = f\,(x,\omega) & \text{in}\;\; D \times \Omega \\
y(x,\omega) = 0 & \text{on}\;\; \partial D \times \Omega
\end{cases}
\tag{3.1}
$$

The following hypotheses on the input data are assumed to hold:

(A1) $a = a(x,\omega) \in L_{\mathbb{P}}^\infty\,(\Omega; L^\infty(D))$ and there exist $a_{min}, a_{max} > 0$ such that

$$
0 < a_{min} \le a(x,\omega) \le a_{max} < \infty \quad \text{a. e. } x \in D \text{ and a. s. } \omega \in \Omega,
$$

(A2) $f \in L_{\mathbb{P}}^2\,(\Omega; L^2(D))$.

Definition 3.1 A random field $y \in L_{\mathbb{P}}^2\,(\Omega; H_0^1(D))$ is said to be a weak solution to (3.1) if it satisfies the so-called variational formulation

$$
\int_\Omega \int_D a\nabla y \cdot \nabla v\, dx d\mathbb{P}(\omega) = \int_\Omega \int_D f v\, dx d\mathbb{P}(\omega) \quad \forall v \in L_{\mathbb{P}}^2\,(\Omega; H_0^1(D)). \tag{3.2}
$$

By considering the bilinear form $A : L_{\mathbb{P}}^2\,(\Omega; H_0^1(D)) \times L_{\mathbb{P}}^2\,(\Omega; H_0^1(D)) \to \mathbb{R}$,

$$
A\,(y,v) = \int_\Omega \int_D a\nabla y \cdot \nabla v\, dx d\mathbb{P}(\omega), \tag{3.3}
$$

and the linear form $F : L_{\mathbb{P}}^2\,(\Omega; H_0^1(D)) \to \mathbb{R}$ defined by

$$
F\,(v) = \int_\Omega \int_D f v\, dx d\mathbb{P}(\omega),
$$

the variational formulation (3.2) is rewritten in the abstract form

$$
A\,(y,v) = F\,(v) \quad \forall v \in L_{\mathbb{P}}^2\,(\Omega; H_0^1(D)).
$$

Thanks to assumption (A1), the bilinear form $A\,(\cdot,\cdot)$ is continuous and elliptic. By (A2), $F\,(\cdot)$ is continuous. As a straightforward application of Lax-Milgram's lemma one has:

Theorem 3.1 *Under assumptions (A1) and (A2) above, the variational problem (3.2) has a unique solution $y = y\,(x,\omega)$ which satisfies the estimate*

$$\|y\|_{L_{\mathbb{P}}^2(\Omega;H_0^1(D))} \le \frac{\dot{C}_P}{a_{min}} \|f\|_{L_{\mathbb{P}}^2(\Omega;L^2(D))}, \tag{3.4}$$

where $C_P = C_P(D)$ is the Poincaré constant, i.e., C_P is the optimal constant in the inequality $\|v\|_{L^2(D)} \le C_P \|\nabla v\|_{L^2(D)}$ for all $v \in H_0^1(D)$.

Inequality (3.4) is easily obtained from asummption (A1) and the Cauchy-Schwarz inequality. Indeed,

$$
\begin{aligned}
a_{min} \int_\Omega \int_D |\nabla y|^2 \, dx d\mathbb{P}(\omega) &\le \int_\Omega \int_D a|\nabla y|^2 \, dx d\mathbb{P}(\omega) \\
&= \int_\Omega \int_D f y \, dx d\mathbb{P}(\omega) \\
&\le \|f\|_{L_{\mathbb{P}}^2(\Omega;L^2(D))} \|y\|_{L_{\mathbb{P}}^2(\Omega;L^2(D))} \\
&\le C_P \|f\|_{L_{\mathbb{P}}^2(\Omega;L^2(D))} \|y\|_{L_{\mathbb{P}}^2(\Omega;H_0^1(D))}.
\end{aligned}
$$

Remark 3.1 Note that due to the isomorphism $L_{\mathbb{P}}^2(\Omega; H_0^1(D)) \cong L_{\mathbb{P}}^2(\Omega) \otimes H_0^1(D)$, a completely analogous analysis of (3.1) may be carried out in the tensor product space $L_{\mathbb{P}}^2(\Omega) \otimes H_0^1(D)$ with the force term $f \in L_{\mathbb{P}}^2(\Omega) \otimes L^2(D)$.

Before proceeding, the following remark on assumptions (A1) and (A2) is in order.

Remark 3.2 As it has been already commented in Sect. 2.3.1, in most applications, Gaussian fields are the model of choice to represent uncertain parameters which show a spatial correlation. Moreover, for numerical simulation purposes, Gaussian fields are typically approximated by truncated KL expansions of the form

$$f(x, \omega) \approx \sum_{n=1}^N \sqrt{\lambda_n} b_n(x) \xi_n(\omega).$$

This choice, which is appropriate for forcing terms and initial conditions, is no longer suitable for random fields which are positive in nature, as the coefficient $a(x, \omega)$ which appears in the principal part of elliptic differential operators. In fact, experimental data (e.g., reported in [15] for the case of steady-state, single-phase groundwater flows) reveal that $a(x, \omega)$ is often log-normal distributed. Hence, as described in Example 2.3, $a(x, \omega)$ is typically approximated as

$$a(x, \omega) \approx e^{\mu(x) + \sigma(x) \sum_{n=1}^N \sqrt{\lambda_n} b_n(x) \hat{\xi}_n(\omega)},$$

where $\hat{\xi}_n$ are truncated standard Gaussian variables.

As a conclusion, assumptions (A1) and (A2) are suitable both from a mathematical point of view and for applications.

Nevertheless, assumptions (A1) and (A2) may be relaxed and still guarantee the well-posedness of (3.2). See [1, Lemma 1.2].

3.1.2   The Heat Equation Revisited I

Let us now study the well-posedness of the parabolic system,

$$\begin{cases} y_t - \text{div}\,(a\nabla y) = 0, & \text{in } (0, T) \times D \times \Omega \\ a\nabla y \cdot n = 0, & \text{on } (0, T) \times \partial D_0 \times \Omega \\ a\nabla y \cdot n = \alpha\,(u - y), & \text{on } (0, T) \times \partial D_1 \times \Omega, \\ y\,(0) = y^0, & \text{in } D \times \Omega, \end{cases} \tag{3.5}$$

which was introduced in Sect. 1.2.2. We recall that in problem (3.5), $D \subset \mathbb{R}^d$ is decomposed into two disjoint parts $\partial D = \partial D_0 \cup \partial D_1$, and n is the unit outward normal vector to ∂D. The random coefficient $a = a(x, \omega)$ satisfies hypothesis (A1) above. In addition, the following conditions are assumed to hold:

(B1) $\alpha = \alpha(x, \omega) \in L^2_{\mathbb{P}}\left(\Omega; L^2(\partial D_1)\right)$, with $\alpha(x, \omega) \geq 0$. There exists a positive constant $\alpha_{max} < \infty$ such that $\alpha(x, \omega) \leq \alpha_{max}$ a. e. $x \in \partial D_1$ and a. s. $\omega \in \Omega$, and

$$\int_\Omega \int_{\partial D_1} \alpha(x, \omega)\,ds d\mathbb{P}(\omega) > 0, \tag{3.6}$$

where ds is the Lebesgue measure on ∂D_1.

(B2) $u = u(t, x, \omega) \in L^2\left(0, T; L^2_{\mathbb{P}}(\Omega; L^2(\partial D_1))\right)$

(B3) $y^0 = y^0(x, \omega) \in L^2_{\mathbb{P}}(\Omega; L^2(D))$.

Let $V = H^1(D)$ and consider the Hilbert space $V_{\mathbb{P}} = L^2_{\mathbb{P}}(\Omega; V)$ equipped with norm

$$\|v\|_{V_{\mathbb{P}}} = \left(\int_\Omega \int_D \left(v^2 + |\nabla v|^2\right) dx\,dP(\omega)\right)^{1/2}.$$

The dual space of $V_{\mathbb{P}}$ is denoted by $V_{\mathbb{P}}^\star$. Finally, consider the space

$$W_{\mathbb{P}}(0, T) = \left\{y \in L^2\,(0, T; V_{\mathbb{P}}) : y_t \in L^2\left(0, T; V_{\mathbb{P}}^\star\right)\right\}$$

endowed with the norm

$$\|y\|_{W_{\mathbb{P}}(0,T)} = \left(\int_0^T \left(\|y(t)\|_{V_{\mathbb{P}}}^2 + \|y_t(t)\|_{V_{\mathbb{P}}^\star}^2\right) dt\right)^{1/2}.$$

Definition 3.2   A random field $y = y(t, x, \omega) \in W_{\mathbb{P}}\,(0, T)$ is a solution to (3.5) if it satisfies

$$\int_0^T (y_t, v)_{V_{\mathbb{P}}^\star, V_{\mathbb{P}}}\,dt + \int_0^T \int_\Omega \int_D a\nabla y \cdot \nabla v\,dx d\mathbb{P}(\omega)dt$$
$$+ \int_0^T \int_\Omega \int_{\partial D_1} \alpha yv\,ds d\mathbb{P}(\omega)dt$$
$$= \int_0^T \int_\Omega \int_{\partial D_1} \alpha uv\,ds d\mathbb{P}(\omega)dt$$

for all $v \in L^2(0, T; V_\mathbb{P})$. Here, $(\cdot, \cdot)_{V_\mathbb{P}^*, V_\mathbb{P}}$ is the duality product[1] between $V_\mathbb{P}$ and $V_\mathbb{P}^*$. The initial condition $y(0, x, \omega) = y^0(x, \omega)$ must be satisfied in $L_\mathbb{P}^2(\Omega; L^2(D))$.

Notice that the space $W_\mathbb{P}(0, T)$ is continuously embedded in $C([0, T]; L_\mathbb{P}^2(\Omega; L^2(D)))$ so that the above initial condition makes sense.

Theorem 3.2 *Let $(\Omega, \mathcal{F}, \mathbb{P})$ be a complete and separable probability space. Let us assume that (A1), (B1)-(B3) hold. Then, there exists a unique solution of (3.5). Moreover, there exists $c > 0$ such that*

$$\|y\|_{W_\mathbb{P}(0,T)} \leq c \left(\|y^0\|_{L_\mathbb{P}^2(\Omega; L^2(D))} + \|u\|_{L^2(0,T; L_\mathbb{P}^2(\Omega; L^2(\partial D_1)))} \right). \qquad (3.7)$$

Proof The proof follows the same lines as in the deterministic case so that here we only focus on the novelty brought about by the probabilistic framework.

We shall apply the abstract result [14, Theorem 26.1]. Both existence and uniqueness are proved as in the deterministic case. Indeed, consider the Hilbert space $H_\mathbb{P} = L_\mathbb{P}^2(\Omega; L^2(D))$ and the Gelfand triple $V_\mathbb{P} \hookrightarrow H_\mathbb{P} \hookrightarrow V_\mathbb{P}^*$. Existence of solutions follows from the Galerkin method. The proof is carried out in several steps:

(i) Since $V_\mathbb{P}$ is isomorphic to $L_\mathbb{P}^2(\Omega) \otimes V$, given the orthonormal bases $\{\phi_i\}_{i \geq 1}$ and $\{\psi_j\}_{j \geq 1}$ of $L_\mathbb{P}^2(\Omega)$ and V, respectively, the set $\{\phi_i \otimes \psi_j\}_{i,j \geq 1}$ is an orthonormal basis of $L_\mathbb{P}^2(\Omega) \otimes V$. Thus, by considering the above orthonormal basis, which for simplicity is denoted by $\{\zeta_i\}_{i \geq 1}$, the solution y is approximated by a sequence

$$y_m(t, x, \omega) = \sum_{i=1}^{m} g_i(t) \zeta_i(x, \omega).$$

(ii) It is proved that $\|y_m\|_{L^2(0,T; V_\mathbb{P})} \leq K$, with K constant. As a consequence, there exists $z = z(t, x, \omega) \in L^2(0, T; V_\mathbb{P})$ such that, up to a subsequence, y_m weakly converges to z. At this point, the crucial step is to prove that the bilinear form $A : V_\mathbb{P} \times V_\mathbb{P} \to \mathbb{R}$ defined as

$$A(y, v) = \int_\Omega \int_D a(x, \omega) \nabla y \cdot \nabla v \, dx d\mathbb{P}(\omega) + \int_\Omega \int_{\partial D_1} \alpha y v \, ds d\mathbb{P}(\omega) \qquad (3.8)$$

is continuous and $V_\mathbb{P}$-elliptic, and that the linear form

$$\begin{aligned} F : V_\mathbb{P} &\to \mathbb{R} \\ v &\mapsto F(v) = \int_\Omega \int_{\partial D_1} \alpha u v \, ds d\mathbb{P}(\omega) \end{aligned} \qquad (3.9)$$

is continuous for a. e. $t \in (0, T)$. The continuity of $A(\cdot, \cdot)$ is an immediate consequence of the Cauchy-Schwarz inequality, the uniform upper bound imposed

---

[1] We recall that for $t \in (0, T)$, $y_t(t) : V_\mathbb{P} \to \mathbb{R}$ is a continuous linear form and $v(t) \in V_\mathbb{P}$. Thus, $(y_t(t), v(t))_{V_\mathbb{P}^*, V_\mathbb{P}} := y_t(t)(v(t))$.

on α in Assumption (B1), and the continuity of the trace operator. $V_{\mathbb{P}}$-ellipticity of $A(\cdot, \cdot)$ is a consequence of (3.6) and may be proved by using an adapted version of the Poincaré inequality, which may be proved by following the same lines as in the deterministic case (see [13, p. 36]). Finally, the continuity of F follows from assumptions (B1)-(B2).

(iii) Finally, it is proved that $z = y$, solution to (3.5).

The rest of the proof is as in the deterministic case (see the proof of [14, Theorem 26.1]).   □

Remark 3.3 Due to the tensor product structure of Bochner spaces, $W_{\mathbb{P}}(0, T)$ may be identified with the spaces

$$W_{\mathbb{P}}(0, T) \cong L_{\mathbb{P}}^2(\Omega) \otimes \left[\left(L^2(0, T) \otimes V\right) \cap \left(H^1(0, T) \otimes V^\star\right)\right]$$
$$\cong L_{\mathbb{P}}^2\left(\Omega; L^2(0, T; V) \cap H^1(0, T; V^\star)\right).$$

Thus, the well-posedness of (3.5) may be studied in these tensor product spaces.

3.1.3 The Bernoulli-Euler Beam Equation Revisited I

To illustrate how the transposition method[2] for proving the well-posedness of second order in time, deterministic, PDEs, extends to random PDEs, we consider the following system for the Bernoulli-Euler beam equation, which was introduced in Sect. 1.2.3:

$$\begin{cases} y_{tt} + (ay_{xx})_{xx} = v\left[\delta_{x_1(\omega)} - \delta_{x_0(\omega)}\right]_x, & \text{in } (0, T) \times D \times \Omega \\ y(t, 0, \omega) = y_{xx}(t, 0, \omega) = 0, & \text{on } (0, T) \times \Omega \\ y(t, L, \omega) = y_{xx}(t, L, \omega) = 0, & \text{on } (0, T) \times \Omega \\ y(0, x, \omega) = y^0(x, \omega), \quad y_t(0, x, \omega) = y^1(x, \omega), & \text{in } D \times \Omega, \end{cases} \quad (3.10)$$

with $D = (0, L)$.

Consider again the Sobolev space $V = H^2(D) \cap H_0^1(D)$ and the Hilbert space $H = L^2(D)$. Consider also the spaces $V_{\mathbb{P}} = L_{\mathbb{P}}^2(\Omega; V)$ and $H_{\mathbb{P}} = L_{\mathbb{P}}^2(\Omega; H)$. The topological dual space of $V_{\mathbb{P}}$ is denoted by $V_{\mathbb{P}}^\star$.

The random coefficient $a = a(x, \omega)$ is assumed to satisfy assumption (A1), as given above. In addition, the following hypotheses on the input data of system (3.10) are assumed to hold:

(C1) $v \in L_{\mathbb{P}}^2\left(\Omega; L^2(0, T)\right)$,
(C2) $0 < x_0(\omega) < x_1(\omega) < L$ for all random events $\omega \in \Omega$, and
(C3) $(y^0, y^1) \in H_{\mathbb{P}} \times V_{\mathbb{P}}^\star$.

---

[2]The concept of transposition lets give a meaning to solutions of PDEs with non-regular input data (initial, boundary conditions and/or forcing terms). To illustrate the main ideas underlying this method, a very simple deterministic example is presented in Sect. 3.4.

Following the same lines as in the case of deterministic PDEs (see [8, Chap. 3]), the concept of a solution for system (3.10) is defined by *transposition* as follows:

- In a first step, for a given $f \in L^2(0, T; H_\mathbb{P})$, consider the problem of finding a function $\zeta = \zeta(t, x, \omega)$ with $\zeta \in L^2(0, T; V_\mathbb{P})$, $\zeta' \in L^2(0, T; H_\mathbb{P})$, $\zeta'' \in L^2(0, T; V_\mathbb{P}^*)$ such that

$$(\zeta'', \varphi)_{V_\mathbb{P}^*, V_\mathbb{P}} + \int_\Omega \int_0^L a(x, \omega) \zeta_{xx} \varphi_{xx} \, dx d\mathbb{P}(\omega) = \int_\Omega \int_0^L f\varphi \, dx d\mathbb{P}(\omega) \quad \forall \varphi \in V_\mathbb{P}, \ (3.11)$$

and a.e. $t \in (0, T)$, where $(\cdot, \cdot)_{V_\mathbb{P}^*, V_\mathbb{P}}$ is the duality product between $V_\mathbb{P}$ and $V_\mathbb{P}^*$. In addition, the final conditions $\zeta(T) = \zeta'(T) = 0$ must be satisfied.
Thanks to hypothesis (A1), the bilinear form $A : V_\mathbb{P} \times V_\mathbb{P} \to \mathbb{R}$, with

$$A(\varphi, \psi) = \int_\Omega \int_0^L a(x, \omega) \varphi_{xx} \psi_{xx} \, dx \mathbb{P}(\omega) \tag{3.12}$$

is continuous and $V_\mathbb{P}$-elliptic. Assuming that $(\Omega, \mathcal{F}, \mathbb{P})$ is separable and following the same lines as in the proof of Theorem 3.2, the Galerkin method may be applied to conclude that there exists a unique solution to (3.11).
- In a second step, the space

$$X_\mathbb{P} = \left\{ \zeta = \zeta_f \text{ solution of (51) with } f \in L^2(0, T; H_\mathbb{P}) \right\},$$

is considered. $X_\mathbb{P}$, endowed with the norm $\|\zeta\|_{X_\mathbb{P}} = \|f\|_{L^2(0,T;H_\mathbb{P})}$, is a Hilbert space. In addition, the operator

$$\zeta \mapsto \zeta'' + (a\zeta_{xx})_{xx}$$

is an isomorphism from $X_\mathbb{P}$ onto $L^2(0, T; H_\mathbb{P})$.
Next, consider the linear and continuous form $\mathcal{L} : X_\mathbb{P} \to \mathbb{R}$ defined by

$$\mathcal{L}(\zeta) = \int_0^T (v[\delta_{x_1} - \delta_{x_0}]_x, \zeta)_{V_\mathbb{P}^*, V_\mathbb{P}} \, dt + (y^1, \zeta(0))_{V_\mathbb{P}, V_\mathbb{P}^*} - \int_\Omega \int_0^L y^0 \zeta'(0) \, dx d\mathbb{P}(\omega).$$

- By transposition [8, Theorem 9.3], there exists a unique $y \in L^2(0, T; H_\mathbb{P})$ such that

$$\int_0^T (y, \zeta'' + (a\zeta_{xx})_{xx})_{H_\mathbb{P}} \, dt = \mathcal{L}(\zeta) \quad \text{for all } \zeta \in X_\mathbb{P}. \tag{3.13}$$

Definition 3.3 The function $y = y(t, x, \omega)$, solution to (3.13) is called a solution to (3.10) in the sense of transposition.

Theorem 3.3 *Let $(\Omega, \mathcal{F}, \mathbb{P})$ be a complete and separable probability space. Let us assume that (A1) and (C1)-(C3) hold. Then, there exists a unique solution y to (3.10) in the sense of transposition. Moreover, y has the regularity*

$$y \in C([0, T]; H_\mathbb{P}) \cap C^1([0, T]; V_\mathbb{P}^*)$$

and there exists c > 0 such that

$$\|y\|_{L^\infty(0,T;H_\mathbb{P})} + \|y'\|_{L^\infty(0,T;V_\mathbb{P}^*)} \le c \left(\|y^0\|_{H_\mathbb{P}} + \|y^1\|_{V_\mathbb{P}^*} + \|v \left[\delta_{x_1} - \delta_{x_0}\right]_x\|_{L^2(0,T;V_\mathbb{P}^*)} \right).$$
$$(3.14)$$

Proof The proof follows exactly the same lines as in [8, Theorem 9.3] so that it is omitted here.   □

3.2 Existence of Optimal Controls Under Uncertainty

In this section, we are concerned with existence of optimal controls for the classes of robust and risk averse control problems introduced in Chap. 1. To illustrate how existence theory of *deterministic* optimal control problems extends to the random case, robust and risk averse control problems for the following Poisson's system are studied next.

With the same notations as in Sect. 3.1.1, let us consider the system

$$\begin{cases} -\text{div}\,(a\nabla y) = 1_{\mathcal{O}}u, & \text{in } D \times \Omega \\ y = 0, & \text{on } \partial D \times \Omega, \end{cases} \quad (3.15)$$

where $y = y(x, \omega)$ is the state variable and $u = u(x)$ is the control variable, which acts on the spatial region $\mathcal{O} \subset D$, a (Lebesgue) measurable subset. As usual, $1_{\mathcal{O}}$ stands for the characteristic function of $\mathcal{O}$.

It is assumed that the random coefficient $a = a(x, \omega)$ satisfies assumption (A1) and that the control $u \in \mathcal{U}_{ad}$, where

$$\mathcal{U}_{ad} = \left\{ u \in L^2(D) : m \le u(x) \le M \quad \text{a.e. } x \in D \right\}, \quad (3.16)$$

with $m, M \in [-\infty, +\infty]$, is the set of admissible controls.

3.2.1 Robust Optimal Control Problems

For a desired target $y_d \in L^2(D)$, the two following cost functionals are considered:

$$J_1(u) = \frac{1}{2} \int_\Omega \int_D |y(x, \omega) - y_d(x)|^2\, dx\, d\mathbb{P}(\omega) + \frac{\gamma}{2} \int_{\mathcal{O}} u^2(x)\, dx, \quad (3.17)$$

where $\gamma \ge 0$ is a weighting parameter, and

$$J_2(u) = \frac{1}{2} \int_D \left(\int_\Omega y(x, \omega)\, d\mathbb{P}(\omega) - y_d(x) \right)^2 dx + \frac{\beta}{2} \int_D \text{Var}\,(y(x))\, dx + \frac{\gamma}{2} \int_{\mathcal{O}} u^2(x)\, dx, \quad (3.18)$$

where

$$\text{Var}\,(y(x)) = \int_\Omega y^2(x,\omega)\,d\mathbb{P}(\omega) - \left(\int_\Omega y(x,\omega)\,d\mathbb{P}(\omega)\right)^2$$

is the variance of $y\,(x,\ldots)$, and $\beta, \gamma \geq 0$ are weighting parameters.

Remark 3.4 Note that tanks to Theorem 3.1, both $J_1\,(u)$ and $J_2\,(u)$ are well-defined for each $u \in \mathcal{U}_{ad}$.

Eventually, consider the two following robust optimal control problems:

$$(P_1) \begin{cases} \text{Minimize in } u \in \mathcal{U}_{ad} : J_1\,(u) \\ \text{subject to} \\ \qquad\qquad y = y\,(u) \quad \text{solves (3.15),} \end{cases}$$

and

$$(P_2) \begin{cases} \text{Minimize in } u \in \mathcal{U}_{ad} : J_2\,(u) \\ \text{subject to} \\ \qquad\qquad y = y\,(u) \quad \text{solves (3.15).} \end{cases}$$

Following the same ideas as in deterministic quadratic optimal control problems [13, Theorem 2.14], the following existence results are established.

Theorem 3.4 *Assume that $\beta = 0$. The following assertions hold:*

(i) *If $-\infty < m < M < +\infty$, then problems (P_1) and (P_2) have, at least, one solution. If, in addition, $\gamma > 0$, then the solution is unique.*

(ii) *If $m = -\infty$ and/or $M = +\infty$, and if $\gamma > 0$, then problems (P_1) and (P_2) have, respectively, a unique solution.*

Proof (i) In the constrained case $(-\infty < m < M < +\infty)$, the set of admissible controls $\mathcal{U}_{ad}$ is bounded, closed and convex. Moreover, by Theorem 3.1, the control-to-state linear operator

$$S : L^2\,(D) \to L^2_\mathbb{P}\,(\Omega; L^2\,(D)), \quad u \mapsto y\,(u)$$

is continuous. Hence, J_1 and J_2 are also continuous. Moreover, J_1 is convex, and, for $\beta = 0$, J_2 is also convex. Consequently, both functionals are weakly lower semi-continuous [2, Chap. 2]. The existence of a solution for (P_1) and (P_2) then follows. The solution is unique if $\gamma > 0$ since, in that case, both J_1 and J_2 are strictly convex.

(ii) In the case $m = -\infty$ and/or $M = +\infty$, the set $\mathcal{U}_{ad}$ is not bounded. However, if $\gamma > 0$, then for a fixed $u^0 \in \mathcal{U}_{ad}$, the search for a solution to (P_1) may be restricted to the closed, convex and bounded set

$$\left\{ u \in L^2\,(D) : \|u\|^2_{L^2(D)} \leq 2\gamma^{-1} J_1\,(u^0) \right\}.$$

Indeed, if $u \in L^2(D)$ satisfies $\|u\|_{L^2(D)}^2 > 2\gamma^{-1} J_1(u^0)$, then

$$J_1(u) \geq \frac{\gamma}{2} \|u\|_{L^2(D)}^2 > J_1(u^0).$$

The same argument applies for J_2. The rest of the proof is the same as in the constrained case.   □

3.2.2 Risk Averse Optimal Control Problems

Let $y_d \in L^2(D)$ be a given target, and $\varepsilon > 0$ a threshold parameter. Consider the cost functional

$$J_\varepsilon(u) = \mathbb{P}\left\{ \omega \in \Omega : I(u, \omega) := \|y(\omega) - y_d\|_{L^2(D)}^2 > \varepsilon \right\}.$$

Notice that since for each $u \in \mathcal{U}_{ad}$, $I(u, \ldots) : \Omega \to \mathbb{R}$ is $\mathcal{F}$-measurable, it defines a real-valued random variable. As a consequence, $\{\omega \in \Omega : I(u, \omega) > \varepsilon\} \in \mathcal{F}$ and thus the cost functional $J_\varepsilon(u)$ is well-defined.

A typical risk averse, optimal control problem is formulated as

$$(P_\varepsilon) \begin{cases} \text{Minimize in } u \in \mathcal{U}_{ad} : J_\varepsilon(u) \\ \text{subject to} \\ \qquad\qquad y = y(u) \quad \text{solves (3.15).} \end{cases}$$

Existence of a solution to (P_ε) is established next.

Theorem 3.5 *Assume that* $-\infty < m < M < +\infty$. *Then, problem* (P_ε) *has, at least, one solution.*

Proof Let u_n be a (bounded) minimizing sequence. Up to a subsequence, still labelled by n, $u_n \rightharpoonup u$ weakly in $L^2(D)$. For a fixed $\omega \in \Omega$, since $I(\cdot, \omega)$ is continuous and convex, it is weakly lower semicontinuous. Hence, $I(u, \omega) \leq \liminf_{n \to \infty} I(u_n, \omega)$. Moreover, the cost functional J_ε may be expressed in integral form as

$$J_\varepsilon(u) = \int_\Omega H(I(u, \omega) - \varepsilon) \, d\mathbb{P}(\omega),$$

where

$$H(s) = \begin{cases} 1, & s > 0 \\ 0, & s \leq 0, \end{cases}$$

is the Heaviside function. Hence, by Fatou's lemma [4, Theorem 9.1],

$$\int_\Omega \liminf_{n\to\infty} H\left(I(u_n,\omega) - \varepsilon\right) d\mathbb{P}(\omega) \le \liminf_{n\to\infty} \int_\Omega H\left(I(u_n,\omega) - \varepsilon\right) d\mathbb{P}(\omega).$$
(3.19)

By the lower semicontinuity of the Heaviside function,

$$H\left(I(u,\omega) - \varepsilon\right) \le \liminf_{n\to\infty} H\left(I(u_n,\omega) - \varepsilon\right)).$$

Thus, integrating in both sides and using (3.19),

$$\int_\Omega H\left(I(u) - \varepsilon\right) d\mathbb{P}(\omega) \le \int_\Omega \liminf_{n\to\infty} H\left(I(u_n) - \varepsilon\right) d\mathbb{P}(\omega)$$
$$\le \liminf_{n\to\infty} \int_\Omega H\left(I(u_n) - \varepsilon\right) d\mathbb{P}(\omega),$$

which completes the proof. $\qquad\square$

Remark 3.5 Existence of optimal controls for (P_ε) in the unconstrained case ($m = -\infty$ and/or $M = +\infty$) mayc be obtained by using the same arguments as in Theorem 3.4.

3.3 Differences Between Robust and Risk-Averse Optimal Control

Although in some situations the optimal control obtained using a risk-averse formulation exhibits an increased robustness, these two approaches show fundamental differences in the control objective and the analysis type.

Regarding the control objective, the robust control approach aims at reducing the system variability under unexpected variations in the input data. However, the risk-averse approach is less concerned with the system variability and more with reducing a risk function that quantifies the expected losses related with the damages caused by extrem events (see Fig. 3.1). The same ideas apply in the context of optimal design problems [5] .

Fig. 3.1 Difference scenarios concerned in robust and risk-averse optimization

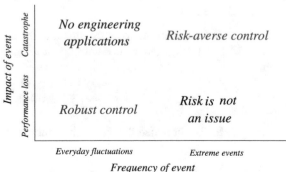

Fig. 3.2 Difference between robustness and risk-aversion

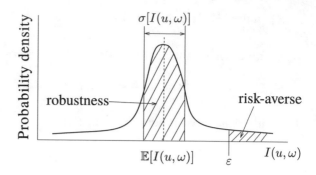

These conceptual differences entail different formulations of the optimal control problem. On the one hand, the robust control formulation incorporates measures of the dispersion of the objective (which for simplicity here and in Fig. 3.2 is denoted by $I(u, \omega)$) around its mean value, such as the standard deviation. On the other hand, the risk-averse approach minimizes a risk function, such as the excess probability, that quantifies the expected losses. These differences in formulation are illustrated in Fig. 3.2. Observe that, whereas in the robust control the attention is paid to the regions around the mean value of the probability density function (PDF) of $I(u, \omega)$, the risk averse approach is focused on the tail of that PDF. As a consequence, since the tails are difficult to approximate numerically, the risk-averse approach involves a higher computational burden compared to its robust counterpart, which only requires the estimation of low order statistical moments (mean and standard deviation).

3.4 Notes

As it has been noticed at the beginning of this chapter, the formalism of tensor product spaces is not necessary to establish an existence theory for solutions of random PDEs. It is however very useful for the numerical approximation of such solutions, as noticed, for instance in [1, 3].

   Existence of solutions for robust and risk averse control problems constrained by evolution random PDEs is obtained by using the same ideas as in the elliptic situation. See, e.g., [9–11].

The Idea of Transposition. To better illustrate the transposition method introduced in Sect. 3.1.3, let us consider the deterministic, non-homogeneous, boundary value problem

$$\begin{cases} -\Delta y = 0 & \text{in} \quad D \\ \quad y = \overline{y} & \text{on} \; \partial D, \end{cases} \tag{3.20}$$

where $D \subset \mathbb{R}^d$ is a bounded domain with a smooth (of class C^2) boundary ∂D, and

$$-\Delta y = -\sum_{j=1}^{d} \frac{\partial^2 y}{\partial x_j^2}$$

is the (negative) of the Laplacian operator acting on y. By using the Lax-Milgram lemma, it is proved that if the boundary datum $\overline{y}$ belongs to the traces space $H^{1/2}(\partial D)$, then there exists a unique weak solution $y \in H^1(D)$ to problem (3.20). However, if $\overline{y} \in L^2(\partial D)$, then the Lax-Milgram lemma can not be applied as in the case where $\overline{y} \in H^{1/2}(\partial D)$. To give a meaning to a solution of (3.20) for $\overline{y} \in L^2(\partial D)$ the transposition method is introduced as follows: let us suppose, for the moment, that $\overline{y}$ is regular and consider the problem

$$\begin{cases} -\Delta z = \varphi & \text{in } D \\ z = 0 & \text{on } \partial D, \end{cases} \tag{3.21}$$

with $\varphi \in L^2(D)$. It is well-known that there is a unique solution $z \in H^2(D)$ to (3.21). Multiplying the PDE in (3.20) by z and integrating by parts twice one gets

$$\int_D y\varphi \, dx = -\int_{\partial D} \overline{y} \frac{\partial z}{\partial n} \, ds. \tag{3.22}$$

The operator $\Lambda : L^2(D) \to H^{1/2}(\partial D)$, defined by $\Lambda \varphi = \frac{\partial z}{\partial n}$, is linear and continuous. When considered as an operator from $L^2(D)$ to $L^2(\partial D)$ it is compact [12, p. 22, Exercice 1.2]. Its adjoint Λ^* is also compact from $L^2(\partial D)$ to $L^2(D)$. Moreover,

$$-\int_{\partial D} \overline{y} \frac{\partial z}{\partial n} \, ds = <\overline{y}, \Lambda\varphi>_{L^2(\partial D)} = <\Lambda^*\overline{y}, \varphi>_{L^2(D)} \quad \forall \varphi \in L^2(D).$$

Thus, by (3.22), $y = \Lambda^*\overline{y}$. Notice that, up to now, to define y we have assumed that $\overline{y} \in H^{1/2}(\partial D)$. However, $\Lambda^*\overline{y}$ is well-defined for $\overline{y} \in L^2(\partial D)$. The transposition method amounts to taking $y = \Lambda^*\overline{y}$ as the solution to (3.20) for $\overline{y} \in L^2(\partial D)$.

This method is of particular importance in the study of control problems associated to that type of PDEs. We refer the reader to [8, Chaps. 2 and 3] for its use in the analysis of PDEs, and to [6, Chap. 4] and [7] for its applications to control of PDEs.

References

1. Babuška, I., Nobile, F., Tempone, R.: A Stochastic collocation method for elliptic partial differential equations with random input data. SIAM Rev. **52**(2), 317–355 (2010)
2. Barbu, V., Precupanu, T.: Convexity and Optimization in Banach Spaces. Springer Monographs in Mathematics, 4 edn. (2012)
3. Gunzburger, M.D., Webster, C., Zhang, G.: Stochastic finite element methods for partial differential equations with random input data. Acta Numer. **23**, 521–650 (2014)
4. Jacod, J., Protter, P.: Probability Essentials, 2nd edn. Springer-Verlag, Berlin, Heidelberg (2003)

5. Kang, Z.: Robust design of structures under uncertainties. Ph.D. Thesis, University of Stuttgart (2005)
6. Lions, J.L.: Optimal Control of Systems Governed by Partial Differential Equations. Springer, Berlin (1971)
7. Lions, J.L., Contröllabilité Exacte, Perturbation et Stabilisation des Systémes Disrtibués. tome 1. Masson (1988)
8. Lions, J.L., Magenes, E.: Non-Homogeneous Boundary Value Problems and Applications Vol. I, Springer-Verlag (1972)
9. Marín, F.J., Martínez-Frutos, J., Periago, F.: Robust averaged control of vibrations for the Bernoulli-Euler beam equation. J. Optim. Theory Appl. **174**(2), 428–454 (2017)
10. Marín, F.J., Martínez-Frutos, J., Periago, F.: A polynomial chaos approach to risk-averse control of stochastic vibrations for the Bernoulli-Euler beam equation. Int. J. Numer. Methods Eng., 1–18 (2018). https://doi.org/10.1002/nme.5823
11. Martínez-Frutos, J., Kessler, M., Münch, A., Periago, F.: Robust optimal Robin boundary control for the transient heat equation with random input data. Internat. J. Numer. Methods Eng. **108**(2), 116–135 (2016)
12. Nečas, J.: Les Méthodes Directes en Théorie des Equations Elliptiques. Éditeurs Academia, Prague (1967)
13. Tröltzsch, F.: Optimal Control of Partial Differential Equations: Theory, Methods and Applications. Graduate Studies in Mathematics 112. AMS. Providence, Rhode Island (2010)
14. Wloka, J.: Partial Differential Equations. Cambridge University Press (1987)
15. Zhang, D.: Stochastic Methods for Flow in Porous Media: Coping with Uncertainties. Academic Press, San Diego, California (2002)

Chapter 4
Numerical Resolution of Robust Optimal Control Problems

> *... a computational solution ... is quite definitely not routine when the number of variables is large. All this may be subsumed under the heading "the curse of dimensionality." Since this is a curse which has hung over the head of the physicist and astronomer for many a year, there is no need to feel discouraged about the possibility of obtaining significant results despite it.*
>
> Richard Bellman.
> Dynamic Programming, 1957.

Both gradient-based methods and methods based on the resolution of first-order optimality conditions may be used for solving numerically the robust optimal control problems presented in the preceding chapters. In both cases, the main difficulty arises in the numerical approximation of statistical quantities of interest associated to solutions of random PDEs. To handle this issue, it is assumed that the random inputs of the underlying PDEs depend on a finite number of random variables. This is the well-known *finite-dimensional noise assumption*. In this chapter, Stochastic Collocation and Stochastic Galerkin methods are presented as powerful tools for solving numerically robust optimal control problems by using gradient-based and one-shot methods.

Complete MatLab codes for all presented examples are available as part of the supporting material accompanying this book.

© The Author(s), under exclusive license to Springer Nature Switzerland AG 2018   45
J. Martínez-Frutos and F. Periago Esparza, *Optimal Control of PDEs under Uncertainty*,
SpringerBriefs in Mathematics, https://doi.org/10.1007/978-3-319-98210-6_4

4.1 Finite-Dimensional Noise Assumption: From Random PDEs to Deterministic PDEs with a Finite-Dimensional Parameter

In many applications, uncertainty in the input data of a PDE is modelled by using a finite number of random variables. In the case of random fields, which typically involve an infinite number of random variables, they are approximated by a finite number of random variables, e.g., with the help of a truncated KL expansion as in (2.12). Random fields which depend on a finite number of random variables are called finite-dimensional noise. All this motivates the following assumption:

> ### Finite dimensional noise assumption
> The random inputs of the PDEs considered from now on depend on a finite number of uncorrelated real-valued random variables
>
> $$\xi(\omega) = (\xi_1(\omega), \cdots, \xi_N(\omega)), \tag{4.1}$$
>
> where $\xi_n : \Omega \to \mathbb{R}$, $1 \le n \le N$.

For clarity in the exposition, from now on in this section we focus on problem (3.1). The following technical result establishes that if the diffusion coefficient $a = a(x, \xi(\omega))$ and the forcing term $f = f(x, \xi(\omega))$ both depend on ξ, then the solution $y = y(x, \omega)$ to (3.1) is also finite-dimensional noise, i.e., $y = y(x, \xi(\omega))$.

Proposition 4.1 *In addition to hypotheses (A1) and (A2), as considered in Sect. 3.1.1, if $a = a(x, \xi(\omega))$ and $f = f(x, \xi(\omega))$, with $\xi(\omega)$ given by (4.1), then the solution y to (3.1) is also finite-dimensional noise, i.e., $y = y(x, \xi(\omega))$.*

The proof of this result is based on the following abstract result, which is specialized to our particular case. For a proof, the reader is referred to [15, pp. 8–9].

Lemma 4.1 (Doob-Dynkin) *Consider the space $(\mathbb{R}^N, \mathscr{B}(\mathbb{R}^N))$, where $\mathscr{B}(\mathbb{R}^N)$ is the σ-algebra of Borel sets in $\mathbb{R}^N$. Let $\xi : \Omega \to \mathbb{R}^N$ be a multivariate random variable and consider the measurable spaces $(\Omega, \sigma(\xi))$, where $\sigma(\xi)$ is the σ-algebra generated by ξ, i.e., $\sigma(\xi) = \{\xi^{-1}(B) : B \in \mathscr{B}(\mathbb{R}^N)\}$. Then, a random variable $g : \Omega \to \mathbb{R}$ is $\sigma(\xi)$-measurable if and only if there exists a measurable function $h : \mathbb{R}^N \to \mathbb{R}$ such that $g = h \circ \xi$.*

Proof of Proposition 4.1. Consider the probability space $(\Omega, \sigma(\xi), \mathbb{P})$. Since $a(x, \cdot)$ and $f(x, \cdot)$ depend on ξ, they are $\sigma(\xi)$-measurable. As a consequence, for every $x \in D$, the solution $y(x, \cdot)$ to (3.1) is also $\sigma(\xi)$-measurable. Indeed, since $a(x, \cdot)$ and $f(x, \cdot)$ are $\sigma(\xi)$-measurable, by Lax-Milgram's lemma, the variational problem: find $y^\star \in L_{\mathbb{P}}^2(\Omega, \sigma(\xi); H_0^1(D))$ such that

$$\int_{\Omega} \int_{D} a\nabla y \cdot \nabla v \, dx d\mathbb{P}(\omega) = \int_{\Omega} \int_{D} f v \, dx d\mathbb{P}(\omega) \quad \forall v \in L^2_{\mathbb{P}}\left(\Omega, \sigma\left(\xi\right); H^1_0\left(D\right)\right)$$

$$(4.2)$$

has a unique solution y^*, which is $\sigma\left(\xi\right)$-measurable as it belongs to the space $L^2_{\mathbb{P}}\left(\Omega, \sigma\left(\xi\right); H^1_0\left(D\right)\right)$. We emphasize that the notation $L^2_{\mathbb{P}}\left(\Omega, \sigma\left(\xi\right); H^1_0\left(D\right)\right)$ indicates that $\sigma\left(\xi\right)$ is the σ-algebra considered in Ω. Considering the same variational problem in the probability space $L^2_{\mathbb{P}}\left(\Omega, \mathscr{F}; H^1_0\left(D\right)\right)$ yields the solution y to problem (3.1). Since $\sigma\left(\xi\right) \subset \mathscr{F}$ is a sub σ-algebra of $\mathscr{F}$, by the uniqueness of solution of problem (4.2), one has $y = y^*$. Hence, $y(x)$ is $\sigma\left(\xi\right)$-measurable for every $x \in D$.

The result then follows by applying Doob-Dynkin's lemma, with $g = y(x)$. $\quad\square$

Remark 4.1 As it will be illustrated hereafter, the main interest of Proposition 4.1 is that it ensures that all statistical information contained in $y(x, \cdot)$ can be obtained from that included in the multivariate random variable ξ.

Remark 4.2 Since material properties and forces are seldom related each other, in practice it is usual to have $a\left(x, \xi\left(\omega\right)\right) = a\left(x, \xi_a\left(\omega\right)\right)$ and $f\left(x, \xi\left(\omega\right)\right) = f\left(x, \xi_f\left(\omega\right)\right)$, with $\xi = \left(\xi_a, \xi_f\right)$ and being ξ_a and ξ_f independent vectors of real-valued random variables.

The abstract probability space $(\Omega, \mathscr{F}, \mathbb{P})$ is not so appealing when statistical quantities associated to solutions of random PDEs must be approximated numerically. Fortunately, under the finite dimensional noise assumption (4.1), one may transform the random PDE (3.1) into a deterministic PDE with a finite dimensional parameter. Indeed, let us denote by $\Gamma_n = \xi_n\left(\Omega\right) \subset \mathbb{R}$ and by $\Gamma = \prod_{n=1}^{N} \Gamma_n \subset \mathbb{R}^N$.

If $g : \Gamma \to \mathbb{R}$ is a Borel measurable function, then the change of variable theorem ([15, p. 21]) yields

$$\int_{\Omega} g\left(\xi\left(\omega\right)\right) d\mathbb{P}\left(\omega\right) = \int_{\Gamma} g\left(z\right) d\mu\left(z\right),$$

where μ is the distribution probability measure of ξ, which is defined on the σ-algebra $\mathscr{B}\left(\Gamma\right)$ of Borel sets on Γ, as

$$\mu\left(B\right) = \mathbb{P}\left(\xi^{-1}\left(B\right)\right), \quad B \in \mathscr{B}\left(\Gamma\right).$$

If, in addition, μ is absolutely continuous with respect to the Lebesgue measure, then the Radon-Nikodym theorem ensures that there exists a joint probability density function for ξ, which from now on is denoted by

$$\rho : \Gamma \to \mathbb{R}_+, \quad \text{with } \rho = \rho\left(z\right) \in L^{\infty}\left(\Gamma\right).$$

Thus,

$$\int_\Omega g\left(\xi\left(\omega\right)\right)d\mathbb{P}\left(\omega\right) = \int_\Gamma g\left(z\right)d\mu\left(z\right) = \int_\Gamma g\left(z\right)\rho\left(z\right)dz,$$

with dz the Lebesgue measure. This way, the abstract probability space $(\Omega, \mathscr{F}, \mathbb{P})$ is mapped to $(\Gamma, \mathscr{B}\left(\Gamma\right), \rho(z)\,dz)$, which is much more suitable for the numerical approximation of statistical quantities of interest associated to solutions of random PDEs.

When applied to problem (3.1), thanks to Proposition 4.1, the variational problem (3.2) rewrites as: find $y = y\left(x, z\right) \in L_\rho^2\left(\Gamma; H_0^1\left(D\right)\right)$ such that

$$\int_\Gamma \rho(z) \int_D a\nabla y \cdot \nabla v\,dxdz = \int_\Gamma \rho(z) \int_D fv\,dxdz \quad \forall v \in L_\rho^2\left(\Gamma; H_0^1\left(D\right)\right),$$
(4.3)

where $L_\rho^2\left(\Gamma; H_0^1\left(D\right)\right)$ denotes the Hilbert space composed of all (equivalence classes of) strongly measurable functions $g : \Gamma \to H_0^1\left(D\right)$ whose norm

$$\|g\|_{L_\rho^2\left(\Gamma; H_0^1(D)\right)} = \left(\int_\Gamma \|g\left(\cdot, z\right)\|_{H_0^1(D)}^2 \rho(z)\,dz\right)^{1/2}$$

is finite. Notice that because of the results of Sect. 2.2 we have the isomorphism

$$L_\rho^2\left(\Gamma; H_0^1\left(D\right)\right) \cong L_\rho^2\left(\Gamma; \mathbb{R}\right) \otimes H_0^1\left(D\right),$$

with $L_\rho^2\left(\Gamma; \mathbb{R}\right) = \left\{g : \Gamma \to \mathbb{R} : \int_\Gamma g^2\left(z\right)\rho(z)\,dz < \infty\right\}$. We emphasize that (4.3) is a *deterministic PDE with a finite-dimensional parameter* $z \in \Gamma$. In its classical form, (4.3) is written as

$$\begin{cases} -\mathrm{div}\left(a(x, z)\nabla y\right) = f, & \text{in } D \times \Gamma \\ y = 0, & \text{on } \partial D \times \Gamma. \end{cases}$$
(4.4)

To sump up, under finite-dimensional noise assumption, a random PDE rewrites as a deterministic PDE with a finite-dimensional parameter.

From now on in this chapter, the finite dimensional noise assumption (4.1) is assumed to hold.

4.2   Gradient-Based Methods

As is well-known [9, Chap. 2], a gradient-based algorithm for the minimization of a generic functional $J : \mathscr{U} \to \mathbb{R}$, with $\mathscr{U}$ a suitable space of admissible controls, is structured as follows:

Gradient-based optimization algorithm

Initialization: Take an initial $u^0 \in \mathcal{U}$.

Construction: For $k = 0, 1, \cdots$ (until convergence) do:

2.1: Choose a descent direction $\overline{u}^k$ (for which $\frac{\partial J(u^k)}{\partial u} \cdot \overline{u}^k < 0$).

2.2: Choose a step size λ^k such that $J\left(u^k + \lambda^k \overline{u}^k\right) < J\left(u^k\right)$.

2.3: Set $u^{k+1} = u^k + \lambda^k \overline{u}^k$.

Verification: The stopping criterion is the first $k \in \mathbb{N}$ for which

$$\frac{|J\left(u^k\right) - J\left(u^{k-1}\right)|}{J\left(u^k\right)} \leq \eta, \tag{4.5}$$

where η is a prescribed tolerance.

In what follows, the chosen descent direction $\overline{u}^k$ is minus the gradient of the cost functional. As for the step size parameter λ^k, it will be chosen as the optimal one in the selected search direction.

Remark 4.3 The stopping criterion (4.5) is often used in practise. Other stopping criteria may also be considered. For instance, criteria based on the size of the descent direction and/or on the relative change in u^k. For simplicity, throughout this text, stopping criteria based on the relative change of the involved cost functionals are used.

4.2.1   Computing Gradients of Functionals Measuring Robustness

The gradients of the cost functionals J_1 and J_2, considered in problems (P_1) and (P_2) of Sect. 3.2.1, are derived by using the formal Lagrange method. The computation of the optimal step size parameters is also indicated next.

Problem (P_1). The cost functional to be minimized is

$$J_1(u) = \frac{1}{2} \int_\Gamma \int_D |y(x, z) - y_d(x)|^2 \, dx \, \rho(z) dz + \frac{\gamma}{2} \int_{\mathscr{O}} u^2(x) \, dx, \tag{4.6}$$

with $\gamma \geq 0$, and y solves

$$\begin{cases} -\mathrm{div}\,(a\nabla y) = 1_{\mathscr{O}} u, & \text{in } D \times \Gamma \\ y = 0, & \text{on } \partial D \times \Gamma. \end{cases} \tag{4.7}$$

For simplicity, it is assumed that no constraints are imposed on the control variable, i.e., $u \in \mathcal{U}_{ad} = L^2(D)$.

To compute such a gradient, the *formal* Lagrange method is used. Although this approach is not mathematically rigorous, it provides a very simple way to derive the optimality conditions of control problems constrained by PDEs. The method is *formal* because it is assumed that all the functions that occur as well as their derivatives are square integrable, without specifying the underlying spaces. We refer the reader to [19, Sects. 2.10 and 2.13] for the basis of the formal Lagrange method for deterministic PDEs. With obvious changes, the method extends to random PDEs, as it is illustrated next.

In a first step, a Lagrange multiplier $p = p(x, z)$ for the random PDE constraint is introduced. Multiplying the PDE in (4.7) by p, integrating by parts, and adding the result to the cost functional (4.6) gives the so-called Lagrangian associated to problem (P_1),

$$\mathscr{L}\left(\hat{y}, \hat{u}, \hat{p}\right) = J_1\left(\hat{y}, \hat{u}\right) - \int_\Gamma \int_D \left[a\nabla\hat{y} \cdot \nabla\hat{p} - 1_{\mathscr{O}}\hat{u}\hat{p}\right] dx\rho(z)dz, \qquad (4.8)$$

where $\hat{y}, \hat{p} \in L^2_\rho\left(\Gamma; H^1_0(D)\right)$, and $\hat{u} \in L^2(D)$. Note that, since in (4.8) the variables $\hat{y}, \hat{u}$ and $\hat{p}$ are treated as independent, we have written $J_1 = J_1\left(\hat{y}, \hat{u}\right)$ and not $J_1\left(\hat{u}\right)$.

Similarly to the case of non-linear mathematical programming problems, we look for a stationary point (y, u, p) of $\mathscr{L}$. Since, formally y is unconstrained, equating the derivative of $\mathscr{L}$ w.r.t. y to zero gives the so-called *adjoint* equation

$$\begin{cases} -\mathrm{div}\,(a\nabla p) = y - y_d, & \text{in } D \times \Gamma \\ p = 0, & \text{on } \partial D \times \Gamma. \end{cases} \qquad (4.9)$$

By inserting the adjoint state p as a test function in the variational formulation of (4.7), the second term in the r.h.s. of (4.8) vanishes. As a consequence, $J_1(u) = \mathscr{L}(y, u, p)$. Thus, $J_1'(u)$ is formally calculated by differentiating the Lagrangian w.r.t. the control u, which leads to

$$J_1'(u)(x) = \gamma u(x) + \int_\Gamma p(x, z)\,\rho(z)dz, \quad x \in \mathscr{O}, \qquad (4.10)$$

where p solves (4.9). As indicated above, a descent direction of the gradient algorithm is given by $\bar{u} = -J_1'(u)$.

Once a descent direction $\bar{u}$ is fixed, the optimal descent parameter λ is obtained by minimizing, over $\mathbb{R}_+$, the function $\lambda \mapsto J_1(u + \lambda\bar{u})$. It is easy to see that $J_1(u + \lambda\bar{u}) = J_1(u) + a_1\lambda + b_1\lambda^2$, where

$$a_1 = \int_\Gamma \int_D \bar{y}(y - y_d)\,dx\rho(z)dz + \gamma\int_{\mathscr{O}} u\bar{u}\,dx, \qquad (4.11)$$

$$b_1 = \frac{1}{2}\int_\Gamma \int_D \bar{y}^2\,dx\rho(z)dz + \frac{\gamma}{2}\int_{\mathscr{O}} \bar{u}^2\,dx \qquad (4.12)$$

and $\bar{y} = \bar{y}(x, z)$ solves

$$\begin{cases} -\text{div}\,(a\nabla\bar{y}) = 1_{\mathcal{O}}\bar{u}, & \text{in } D \times \Gamma \\ \bar{y} = 0, & \text{on } \partial D \times \Gamma. \end{cases}$$

Hence, the optimal step size parameter is $\lambda_1 = -\frac{a_1}{2b_1}$.

Problem $(\mathbf{P_2})$. Consider the cost functional $J_2(u)$, as given by (3.18), which now is expressed as

$$J_2(u) = \frac{1}{2}\int_D \left(\int_\Gamma y(x, z)\rho(z)\,dz - y_d(x)\right)^2 dx + \frac{\beta}{2}\int_D \text{Var}\,(y(x))\,dx + \frac{\gamma}{2}\int_{\mathcal{O}} u^2(x)\,dx, \tag{4.13}$$

Proceeding as in the preceding case,

$$J_2'(u)(x) = \gamma u(x) + \int_\Gamma q(x, z)\,\rho(z) \quad x \in \mathcal{O}, \tag{4.14}$$

where $q = q(x, z)$ solves the adjoint system

$$\begin{cases} -\text{div}\,(a\nabla q) = \int_\Gamma y\,\rho(z)dz - y_d + \beta\left(y - \int_\Gamma y\,\rho(z)dz\right), & \text{in } D \times \Gamma \\ q = 0, & \text{on } \partial D \times \Gamma. \end{cases} \tag{4.15}$$

The descent direction is $\hat{u} = -J_2'(u)$. Similarly to the preceding case, the optimal step size parameter is $\lambda_2 = -\frac{a_2}{2b_2}$, where

$$\begin{aligned} a_2 = &\int_D \left(\int_\Gamma y\,\rho(z)dz - y_d\right)\left(\int_\Gamma \hat{y}\,\rho(z)dz\right) dx \\ &+\beta\int_D \left[\int_\Gamma y\hat{y}\,\rho(z)dz - \int_\Gamma y\,\rho(z)dz\int_\Gamma \hat{y}\,\rho(z)dz\right] dx \\ &+\gamma\int_{\mathcal{O}} u\hat{u}\,dx, \end{aligned} \tag{4.16}$$

$$b_2 = \frac{1}{2}\int_D \left(\int_\Gamma \hat{y}\,\rho(z)dz\right)^2 dx + \frac{\beta}{2}\int_D \text{Var}\,(\hat{y}(x))\,dx + \frac{\gamma}{2}\int_{\mathcal{O}} \hat{u}^2\,dx, \tag{4.17}$$

and $\hat{y}(x, z)$ is a solution to

$$\begin{cases} -\text{div}\,(a\nabla\hat{y}) = 1_{\mathcal{O}}\hat{u}, & \text{in } D \times \Gamma \\ \hat{y} = 0, & \text{on } \partial D \times \Gamma. \end{cases}$$

As is observed in both problems (P_1) and (P_2), the main task to implement a gradient-based optimization algorithm arises in the numerical approximation of statistical quantities (mean and variance in the case of robust control problems) associated to solutions of random PDEs. This crucial fact is addressed in the next section.

4.2.2 Numerical Approximation of Quantities of Interest in Robust Optimal Control Problems

The following reference benchmark problem is considered again

$$\begin{cases} -\text{div}\,(a\,(x,z)\,\nabla y\,(x,z)) = 1_{\mathcal{O}} u\,(x)\,, & (x,z) \in D \times \Gamma \\ y\,(x,z) = 0, & (x,z) \in \partial D \times \Gamma, \end{cases} \tag{4.18}$$

where $\Gamma \subset \mathbb{R}^N$ is a compact set. As it has been showed above, in robust optimal control problems one is mainly interested in computing the first and second order statistical moments of the solution $y\,(x,z)$,

$$\mathbb{E}\left[y^j\right](x) = \int_\Gamma y^j\,(x,z)\,\rho(z)dz, \quad j = 1, 2, \quad x \in D. \tag{4.19}$$

A number of methods (e.g., Monte Carlo (MC), Stochastic Galerkin (SG), Stochastic Collocation (SC), Reduced Basis (RB), etc.) may be used to approximate numerically (4.19). Although MC sampling is a natural choice, its very slow convergence may lead to an unaffordable computational cost when used in optimization algorithms that require iteration. In addition, MC method does not exploit the possible regularity of the solution w.r.t. the random parameter. For the specific problems under consideration in this chapter, the solution y of (4.18) is *smooth* w.r.t. the parameter $z \in \Gamma$. Indeed, it is known [2] that, under reasonable assumptions on the uncertain coefficient $a\,(x,z)$, the solution y to (4.18) is analytic w.r.t. the parameter z. More details on this issue are provided in Sect. 4.5 below. In this situation, stochastic collocation methods are computationally very efficient, at least when the number N of involved random variables is not so large, as is the case of truncated KL expansions of smooth random fields. Accordingly, an adaptive, anisotropic, sparse grid, stochastic collocation method is used to compute the statistical quantities involved in problems (P_1) and (P_2).

4.2.2.1   An Anisotropic, Sparse Grid, Stochastic Collocation Method

Throughout this section, we closely follow [2, 12] for the description of the method and [3] for its numerical implementation. We focus on an algorithmic presentation of stochastic collocation methods and refer the readers to [2, 12] for error estimates. See also [17, Chap. 11].

The basic idea to approximate (4.19) is to use a quadrature rule so that the solution $y\,(x,z)$ to (4.18) must be known at a set of nodes $z^k = \left(z_1^k, \cdots, z_N^k\right) \in \Gamma$, $1 \le k \le M$, as well as suitable weights w^k, $1 \le k \le M$. The choice of these nodes and weights is a crucial point as it determines the accuracy and efficiency of the collocation method. Although the first idea would be to use a full tensor product of a one dimensional set, even for a reduced number of random variables, full tensor rules are, in most of the cases of interest, unnecessary, in the sense that the same exactness may be achieved by using a lower number of nodes. Sparse grid-based quadrature rules must be used instead. Both, isotropic and anisotropic quadrature rules may be considered. Having in mind that when random inputs are represented by truncated KL expansions of random fields, the different stochastic directions do not weigh equally, anisotropic rules may preserve similar accuracy as isotropic ones, but alleviating the involved computational cost.

For pedagogical reasons, in parallel to the theoretical description of the method, the numerical approximation of (4.19) is illustrated through the following specific data: $D = (0, 1)^2$ is the unit square, $\mathcal{O}$ is a circle centred at $(0.5, 0.5)$ and of radius 0.25, and $u(x) = 1$ for all $x \in \mathcal{O}$. The random domain is $\Gamma = [-3, 3]^N$, and $a(x, z)$ is the truncated KL expansion considered in (2.29), with $N = 4$, i.e.,

$$a(x, z) = e^{\mu + \sigma \sum_{n=1}^{4} \sqrt{\lambda_n} b_n(x) z_n}, \quad x = (x_1, x_2) \in D, \quad z = (z_1, z_2, z_3, z_4) \in \Gamma, \tag{4.20}$$

where $\mu = -0.3466$ and $\sigma = 0.8326$; $\{\lambda_j, b_j(x)\}_{j=1}^{4}$ are the first fourth eigenpairs associated to the covariance function (2.28), with correlation lengths $L_1 = L_2 = 0.5$, and finally

$$\rho(z) = \prod_{n=1}^{4} \phi(z_n), \quad z = (z_1, z_2, z_3, z_4) \in \Gamma, \tag{4.21}$$

where $\phi(z_n)$ is the truncated Gaussian (2.30). See Example 2.3 for more details.

To better understand an anisotropic, sparse grid, quadrature rule in the multi-dimensional case $N > 1$, let us start with the one dimensional case $N = 1$.

The one-dimensional case. Let $f : \Gamma_1 \subset \mathbb{R} \to \mathbb{R}$ be a random function and $\rho_1 = \rho_1(z_1)$ the PDF associated to the random parameter $z_1 \in \Gamma_1$. It is useful to arrange the number of quadrature nodes in levels. Thus, for a positive integer $\ell \in \mathbb{N}_+$, from now on called the *level* of the quadrature rule, the number of nodes in the level ℓ is denoted by m_ℓ. Several choices are possible to define m_ℓ. For instance, one may add just only one quadrature node when passing from one level to the next one, or one may double the number of nodes. Another typical choice of m_ℓ is [18]

$$m_1 = 1 \quad \text{and} \quad m_\ell = 2^{\ell-1} + 1 \text{ for } \ell > 1. \tag{4.22}$$

A quadrature rule for f at the level ℓ then takes the form

$$\int_{\Gamma_1} f(z_1) \rho_1(z_1) \, dz_1 \approx \sum_{j=1}^{m_\ell} f\left(z_1^j\right) w_1^j := \mathcal{Q}^\ell f, \tag{4.23}$$

where z_1^j and w_1^j are, respectively, the nodes and weights of the one-dimensional quadrature rule.

Selection of quadrature nodes and weights. Several choices are possible for the nodes z_1^j and the weights w_1^j. These are typically classified into nested and non-nested quadrature rules. The former are characterized by the fact that the nodes at the level ℓ contain the nodes at the level $\ell - 1$. This property does not hold in non-nested rules. Clenshaw-Curtis abscissas, in which the nodes are the extrema of Chebyshev polynomials, which are given by

$$z_1^j = -\cos\left(\frac{\pi(j-1)}{m_\ell - 1}\right), \quad j = 1, \cdots, m_\ell, \tag{4.24}$$

are widely used as nested quadrature rules.

Gaussian abscissas, in which the nodes are the zeros of a suitable class of orthogonal polynomials (e.g., Hermite polynomials for the case of Gaussian random variables, and Legendre polynomials for uniformly distributed random variables) are the classical examples of non-nested quadrature rules. Gauss-Hermite nodes are used in the numerical experiments included in this chapter. The associated weights are those for which the quadrature rule (4.23) is exact for polynomials up to degree $m_\ell - 1$.

For a comparison of Clenshaw-Curtis and Gaussian quadrature rules we refer to [20].

The multi-dimensional case $N > 1$. Let $y(x, \cdot) : \Gamma \subset \mathbb{R}^N \to \mathbb{R}$ be a solution to (4.18) and let us assume that $y(x, z)$ admits an analytical extension w.r.t. each stochastic direction z_n, $1 \leq n \leq N$, in a region of the complex plane $\mathbb{C}$. As it will be indicated in Sect. 4.5 below, the size of these regions depends on $\sqrt{\lambda_n}\|b_n\|_{L^\infty(D)}$. Recall that throughout this text, hypotheses of Mercer's Theorem 2.2 are assumed to hold. As a consequence, the eigenfunctions $b_n \in L^\infty(D)$. Stochastic directions in which the region of analyticity of the solution is larger require less collocation points than directions with less regularity. Accordingly, a vector of weights $g = (g_1, \cdots, g_N)$ for the different stochastic directions is introduced. Vector g will determine the number of collocation points to be used in each stochastic direction. For the specific problem under consideration, in which $a(x, z)$ is modelled by a truncated KL expansions of a log-normal field, it is proved in [2] that an appropriate selection of g is

$$g_n = \frac{1}{2\sqrt{2\lambda_n}\|b_n\|_{L^\infty(D)}}, \quad 1 \leq n \leq N. \tag{4.25}$$

Thus, the larger the $\sqrt{\lambda_n}\|b_n\|_{L^\infty(D)}$ (or equivalently the smaller the g_n) the more important the corresponding n-th stochastic direction and so the larger the number of collocation points required in that stochastic direction (see [2, 12] for more details on this passage).

For the specific case of the random field (4.20) one has

$$g = (0.3894, 0.5478, 0.5478, 0.7708). \tag{4.26}$$

Starting from a fixed quadrature level $\ell \in \mathbb{N}_+$ for the multi-dimensional grid, and from the vector of weights g, in order to construct an anisotropic tensor grid, it is very convenient to use a multi-index notation. Hence, the following multi-index set is introduced:

$$\mathbb{I}_g\left(\ell, N\right) = \left\{ i = (i_1, \cdots, i_N) \in \mathbb{N}_+^N, \quad i \geq 1 : \quad \sum_{n=1}^{N} (i_n - 1) g_n \leq \ell \underline{g} \right\}, \quad (4.27)$$

where $\underline{g} = \min_{1 \leq n \leq N} g_n$. Note that the lower the weight g_n, the larger the corresponding maximum index in the n-th direction. Also note that (4.27) is completely determined by the weighting vector g and the level ℓ. This indices set together with the rule used to increase the number of points in each direction (i.e., increasing by one point to pass from one level to the next one, doubling the number of points, using (4.22), etc.), determines the total number of collocation points in the multi-dimensional grid. Since the sparse grid generated may have points in common, in order to save computational time during evaluation of a function on a sparse grid, it is then important to get ride of these repetitions.

To construct the anisotropic quadrature rule associated to the multi-index set $\mathbb{I}_g\left(\ell, N\right)$, let us first consider the difference operator

$$\Delta^j = \mathcal{Q}^j - \mathcal{Q}^{j-1}, \quad \text{for } j = 1, 2, \ldots \quad (4.28)$$

with $\mathcal{Q}^0 = 0$, and where $\mathcal{Q}^j$ is the one-dimensional quadrature rule given by (4.23).

By using (4.27) and (4.28), the so-called anisotropic Smolyak[1] quadrature rule approximating $\int_\Gamma y\left(x, z\right) \rho(z) dz$, is defined as

$$\mathcal{A}_g\left(\ell, N\right) y\left(x\right) := \sum_{i \in \mathbb{I}_g(\ell, N)} \left(\Delta^{i_1} \otimes \cdots \otimes \Delta^{i_N}\right) y\left(x\right), \quad (4.29)$$

which is equivalent to

$$\mathcal{A}_g\left(\ell, N\right) y\left(x\right) = \sum_{i \in Y_g(\ell, N)} c_g(i) \left(\mathcal{Q}^{i_1} \otimes \cdots \otimes \mathcal{Q}^{i_N}\right) y\left(x\right), \quad (4.30)$$

with

$$c_g(i) = \sum_{\substack{j \in \{0, 1\}^N \\ i + j \in \mathbb{I}_g\left(\ell, N\right)}} (-1)^{|j|}, \quad |j| = \sum_{k=1}^{N} j_k \quad (4.31)$$

---

[1]The sparse grid quadrature rule presented in this chapter was proposed by Smolyak [18]. The main idea underlying this construction is described in Sect. 4.5.

and

$$Y_g\left(\ell, N\right) = \mathbb{I}_g\left(\ell, N\right) \setminus \mathbb{I}_g\left(\ell - \frac{|g|}{\underline{g}}, N\right), \quad |g| = \sum_{k=1}^{N} g_k. \qquad (4.32)$$

Thus, to compute $\mathscr{A}_g\left(\ell, N\right) y\left(x\right)$, we only need to determine the coefficients $c_g(i)$ and to apply tensor product quadrature formulae. We recall that for a fixed multi-index $i = (i_1, \cdots, i_N) \in Y_g\left(\ell, N\right)$,

$$\left(\mathscr{Q}^{i_1} \otimes \cdots \otimes \mathscr{Q}^{i_N}\right) y\left(x\right) := \sum_{r_1=1}^{m_{i_1}} \cdots \sum_{r_N=1}^{m_{i_N}} y\left(x; z_1^{r_1}, \cdots, z_N^{r_N}\right) w_{i_1}^{r_1} \cdots w_{i_N}^{r_N}, \qquad (4.33)$$

where $\left(z_j^{r_j}, w_{i_j}^{r_j}\right)$, $1 \leq j \leq N$, $1 \leq r_j \leq m_{i_j}$, are the one-dimensional nodes and weights associated to the level m_{i_j}. Thus, by considering the anisotropic nodal set

$$\Theta_g\left(\ell, N\right) = \bigcup_{i \in Y_g(\ell, N)} \left(\Theta^{i_1} \times \cdots \times \Theta^{i_N}\right), \qquad (4.34)$$

with Θ^{i_n} the set of one-dimensional nodes for the level i_n in the n-th stochastic direction, (4.30) may be expressed in the form

$$\mathscr{A}_g\left(\ell, N\right) y\left(x\right) = \sum_{z^k \in \Theta_g(\ell, N)} y\left(x, z^k\right) w^k, \qquad (4.35)$$

where the weights w^k are obtained by multiplying the corresponding coefficients $c_g(i)$ by its associated weights, which appear in (4.33). Notice that, as a consequence, some weights w^k may take negative values.

Remark 4.4 Having in mind an efficient numerical implementation of the multivariate quadrature rule (4.35), it is convenient to compute firstly the coefficients (4.31) because some of those coefficients may be equal to zero. Hence, there is no need to compute nodes and weights associated to indices $i \in Y_g\left(\ell, N\right)$ for which $c_g(i) = 0$.

Remark 4.5 It is proved in [13] that for a multivariate function $f = f(z_1, \cdots, z_N)$ having continuous derivatives up to order α in each variable, if we denote by M the number of nodes in the sparse grid, then the quadrature error is of order $O\left(M^{-\alpha} (\log M)^{(N-1)(\alpha+1)}\right)$.

As an example, Table 4.1 collects the maximum index in the n-th stochastic direction (which is denoted by β_n,) obtained according to (4.27), for the vector of weights (4.26) and for several levels ℓ. One-dimensional Gauss-Hermite nodes are ordered in levels m_{i_n}, with $m_{i_n} = i_n$. The total number of nodes, after removing repeated nodes and nodes associated to coefficients $c_g(i)$ which are equal to zero, is shown in the last column.

Table 4.1 Maximum index set for the vector of weights (4.26) and for several levels ℓ

ℓ	β_1	β_2	β_3	β_4	# Nodes
1	2	1	1	1	2
2	3	2	2	2	9
3	4	3	3	2	31
4	5	3	3	3	61
5	6	4	4	3	149
8	10	7	7	5	1851

Example 4.1 To better illustrate the construction of the above anisotropic, sparse grid, and to show the differences between isotropic and anisotropic sparse grids, Fig. 4.1 displays the multi-index sets $\mathbb{I}_g$ (ℓ, N) for two stochastic directions ($N = 2$), and the vectors of weights $g = (1, 1)$ (isotropic) and $g = (1, 0.5)$ (anisotropic), both for a quadrature level $\ell = 4$. The associated isotropic and anisotropic sparse grids are composed of Gaussian nodes (zeros of Hermite polynomials). The number of one-dimensional nodes m_{i_n}, $n = 1, 2$, in the n-th stochastic direction is taken as $m_{i_n} = i_n$. Thus, the nodes corresponding to the one-dimensional level i_n are the zeros of the Hermite polynomial of degree i_n.

To construct the multi-index set $Y_{(1,0.5)}$ $(4, 2)$ (as given by (4.32)) in the anisotropic case (Fig. 4.1c–d), we only have to remove the indices $(1, 1)$ and $(1, 2)$ from Fig. 4.1c. From (4.31) it follows that $c_{(1,4)}$ $(1, 0.5) = c_{(1,0.5)}$ $(2, 2) = 0$. Consequently, the nodes associated to the indices $(1, 4)$ and $(2, 2)$ are not plotted in Fig. 4.1d.

Practical numerical implementation. Discretization in the physical domain.
Once the set of nodes Θ_g (ℓ, N) and it associated weights have been computed, to approximate the first two statistical moments, $\mathbb{E}\left[y^j\right](x)$, $j = 1, 2$, of the solution $y(x, z)$ to (4.18) we proceed as follows:

(i) Compute, by using finite elements, the solutions of the deterministic problems

$$\begin{cases} -\text{div}\left(a\left(x, z^k\right) \nabla y\left(x, z^k\right)\right) = 1_\mathcal{O} u(x), & x \in D \\ y\left(x, z^k\right) = 0, & x \in \partial D, \end{cases} \quad (4.36)$$

at each node $z^k \in \Theta_g$ (ℓ, N).
(ii) Apply the quadrature rule (4.35) to $y(x^j)$ (and to $y^2(x^j)$) at each node x^j of the finite element mesh in the physical domain D.

Figure 4.2 displays the expected value $\mathbb{E}[y](x)$ (left) and the variance $\text{Var}[y](x) = \mathbb{E}\left[y^2\right](x) - (\mathbb{E}[y](x))^2$ (right) of the solution to (4.18) for the data: $D = (0, 1)^2$, $\mathcal{O}$ a circle centred at $(0.5, 0.5)$ and of radius 0.25, and $u(x) = 1$ for all $x \in \mathcal{O}$. The random domain $\Gamma = [-3, 3]^N$, and $a(x, z)$ is given by (4.20). Lagrange P_1 finite elements have been used for approximation in the physical domain D. The mesh used for discretization of D is plotted in Fig. 4.3. Approximation in the random domain Γ has been carried out by using the anisotropic sparse grid collocation method described in the preceding section. The level ℓ of the multivariate quadrature rule has been fixed to $\ell = 5$. Gauss-Hermite nodes and weights have been used.

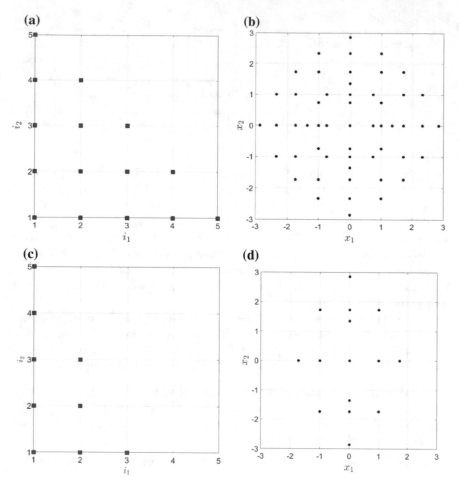

Fig. 4.1 $N = 2$, $\ell = 4$. (**a**) Multi-index set $\mathbb{I}_g$ (ℓ, N), with $g = (1, 1)$ and associated isotropic sparse grid (**b**). (**c**) Multi-index set $\mathbb{I}_g$ (ℓ, N) for $g = (1, 0.5)$ and associated anisotropic sparse grid (**d**). Gaussian nodes corresponding to zeros of Hermite polynomials are plotted in (**b**) and (**d**). Both the indices sets and sparse grids have been generated by using the Sparse grids Matlab kit [3] (http://csqi.epfl.ch)

4.2.2.2   Adaptive Algorithm to Select the Quadrature Level ℓ

This subsection describes an adaptive algorithm to select the level ℓ that will be used, along the descent algorithm, to approximate (in the random domain) the statistical quantities of interest in problems (P_1) and (P_2).

Since the main goals in (P_1) and (P_2) are to minimize the cost functionals $J_1(u)$ and $J_2(u)$, the level ℓ used for numerical integration in the random domain is adaptively chosen so as to comply with a prescribed accuracy level δ. Obviously, δ should be less than or equal to the tolerance imposed in the stopping criterion (4.5) of the

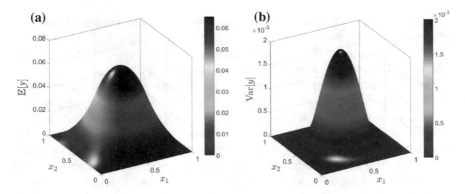

Fig. 4.2 Approximation of the expected value $\mathbb{E}[y](x)$ **(a)** and the variance $\text{Var}[y](x)$ **(b)** for problem (4.18)

Fig. 4.3 Mesh of the physical domain D used for finite elements approximation

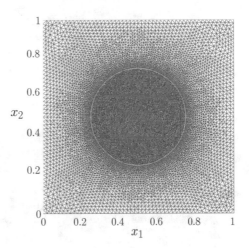

gradient algorithm, i.e., $\delta \leq \eta$. The idea is therefore quite simple: roughly speaking, firstly, a criterion to measure the approximation error of the statistical quantities of interest is considered. Then, the level ℓ is increased linearly up to that criterion is satisfied.

To fix ideas, let us consider problem (P_2), where the cost functional $J_2(u)$, given by (4.13), involves the numerical approximation of the first two statistical moments considered in (4.19). Let us denote by $\mathcal{M}_{g,\ell,N}^{j,h}(y)$ a numerical approximation of

$$\int_D \int_\Gamma y^j(x,z)\, \rho(z)\, dzdx, \quad j = 1, 2, \tag{4.37}$$

obtained as described in the preceding subsection. Here, h refers to the size of the finite element mesh for discretization of the physical domain D, and g, ℓ, N are the parameters that appear in (4.29).

Since the level ℓ is going to be fixed before running the optimization algorithm, a reference control function $u = u(x)$, which appears in the right-hand side of (4.18), must be fixed in advance. For instance, one may consider the optimal control of the associated deterministic control problem (see Sect. 4.2.3.1 for a specific example).

Then, to estimate the computational error in the level ℓ, reference (or enriched) values for (4.37) are computed before optimization. This could be done, e.g., by using Monte Carlo method. If the computational cost of MC method is unaffordable, then, following [12], one may use the following approach: take a large enough integer $\bar{\ell}$. Denoting by $\beta_n(g, \bar{\ell}) := \max_{i \in \mathbb{I}_g(\bar{\ell}, N)}\{i_n\}$ the maximum index, in the n-th stochastic direction, used in the quadrature rule (4.29), enriched (or reference) values for (4.37) are obtained by using the quadrature rule (4.29) for the level $\bar{\ell} + 1$ and for a larger vector of weights $\hat{g}$, which have to satisfy that $\beta_n(\hat{g}, \bar{\ell} + 1) \geq \beta_n(g, \bar{\ell}) + 1$. As indicated in [12], $\hat{g}$ may be defined component-wise as

$$\hat{g}_n = \left(\frac{\beta_n(g, \bar{\ell}) - 1}{\beta_n(g, \bar{\ell})}\right)\left(\frac{g_n}{g}\right).$$

These enriched values are denoted by $\mathcal{M}^{j,h}_{\hat{g},\bar{\ell}+1,N}(y)$, $j = 1, 2$, and play the role of exact computations for (4.37).

To sum up, the adaptive algorithm to select ℓ is the following:

Adaptive algorithm to select the level ℓ of quadrature
1. Fix a tolerance $0 < \delta \ll 1$ and an enriched level of quadrature $\bar{\ell}$.
2. Compute enriched values, $\mathcal{M}^{j,h}_{\hat{g},\bar{\ell}+1,N}(y)$, for (4.37).
3. Increasing the level ℓ linearly from $\ell = 1$ to $\ell = \ell_{opt} \leq \bar{\ell}$ until the following stopping criterion is satisfied

$$\max\left\{\frac{|\mathcal{M}^{1,h}_{\hat{g},\bar{\ell}+1,N}(y) - \mathcal{M}^{1,h}_{g,\ell,N}(y)|}{\mathcal{M}^{1,h}_{\hat{g},\bar{\ell}+1,N}(y)}, \frac{|\mathcal{M}^{2,h}_{\hat{g},\bar{\ell}+1,N}(y) - \mathcal{M}^{2,h}_{g,\ell,N}(y)|}{\mathcal{M}^{2,h}_{\hat{g},\bar{\ell}+1,N}(y)}\right\} \leq \delta.$$
(4.38)

If the stopping criterion (4.38) is not satisfied, then it means that the integer $\bar{\ell}$ is not large enough as to approximate with enough accuracy the exact quantity of interest. In such a case, another $\bar{\ell}$ greater than the previous one is selected and the whole process is repeated.

Remark 4.6 For the case of problem (P_1), denoting by $\mathcal{M}^h_{g,\ell,N}$ a numerical approximation of

$$\int_\Gamma \int_D |y(x, z) - y_d(x)|^2 dx\rho(z)dz$$
(4.39)

the adaptive algorithm described above applies, but with the stopping criterion (4.38) replaced by

$$\frac{|\mathcal{M}^h_{\hat{g},\overline{\ell}+1,N}(y) - \mathcal{M}^h_{\hat{g},\ell,N}(y)|}{\mathcal{M}^h_{\hat{g},\overline{\ell}+1,N}(y)} \leq \delta, \tag{4.40}$$

where, similarly to problem (P_2), $\mathcal{M}^h_{\hat{g},\overline{\ell}+1,N}$ denotes an enriched numerical approximation of (4.39).

Remark 4.7 Convergence results for this type of adaptive algorithms, in terms of the level ℓ, have been obtained in [2, Sect. 4.1] and in [12, Theorem 3.13].

Figures 4.6 and 4.11 illustrate, in two specific examples, the performance of the adaptive algorithm described in this subsection.

4.2.3 Numerical Experiments

This section presents numerical simulation results for problems (P_1) and (P_2) obtained by applying the gradient algorithm described in Sect. 4.2, in combination with the above adaptive, anisotropic, sparse grid, stochastic, collocation method.

As in the preceding section, in what follows, it is assumed that the physical domain $D = (0, 1)^2$ is the unit square and the control region

$$\mathcal{O} = \left\{ x = (x_1, x_2) \in \mathbb{R}^2 : (x_1 - 0.5)^2 + (x_2 - 0.5)^2 < 0.25^2 \right\},$$

is a circle centred at the point $(0.5, 0.5)$ and of radius $r = 0.25$. The random coefficient $a(x, z)$ is the truncated KL expansion considered in (4.20), the random domain is $\Gamma = [-3, 3]^4$, and $\rho(z)$ is given by (4.21).

4.2.3.1 A Deterministic Test Example

In order to assess and compare numerical simulation results for optimal control problems under uncertainty with their deterministic counterparts, the following deterministic optimal control problem is introduced next.

Consider the control function $u(x) = 1, x \in \mathcal{O}$, and let $y_d = y_d(x)$ be the solution of the deterministic problem

$$\begin{cases} -\text{div}(\nabla y) = 1_{\mathcal{O}} u, & \text{in } D, \\ y = 0, & \text{on } \partial D. \end{cases} \tag{4.41}$$

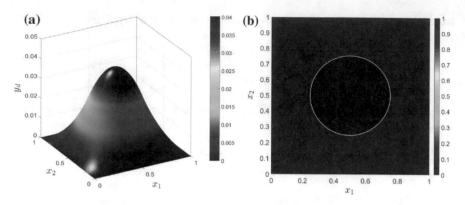

Fig. 4.4 (a) Target function y_d. (b) Optimal exact control u_{exact}

Then, obviously, the deterministic optimal control problem

$$
(P_{det})
\begin{cases}
\text{Minimize in } u \in L^2(D) : J_{det}(u) = \frac{1}{2} \int_D |y(x) - y_d(x)|^2 dx \\
\text{subject to} \\
\qquad\qquad -\text{div}(\nabla y) = 1_{\mathscr{O}} u, \quad \text{in } D, \\
\qquad\qquad\qquad\; y = 0, \qquad\quad\;\; \text{on } \partial D,
\end{cases}
\tag{4.42}
$$

has the solution $u_{exact}(x) = 1, x \in \mathscr{O}$.

Figure 4.4 displays the target function $y_d(x)$ and the optimal exact control u_{exact}.
Problem (P_{det}) has been solved by using a standard gradient algorithm with optimal step-size. Numerical simulation results are plotted in Fig. 4.5. Precisely, Fig. 4.5a displays the error with respect to the target, i.e., the values of

$$
|y(u_{det}) - y(u_{exact})|
$$

at each point $x \in D$, where u_{det} is the numerical solution of (P_{det}). Here, $y(u_{exact})$ and $y(u_{det})$ denote, respectively, the solutions to (4.18), with $u = u_{exact}$ and $u = u_{det}$. Figure 4.5b shows the optimal numerical control u_{det}.

Figure 4.7a displays the convergence history for problem (P_{det}).

4.2.3.2   Experiment 1: Problem (P_1)

Next, we solve (P_1) for the case $\gamma = 0$ in (4.6). The target function $y_d(x)$ is the same as in the deterministic problem, i.e., the function y_d is the solution to (4.41) for the control $u(x) = 1, x \in \mathscr{O}$.

The algorithm to solve (P_1) is the following:

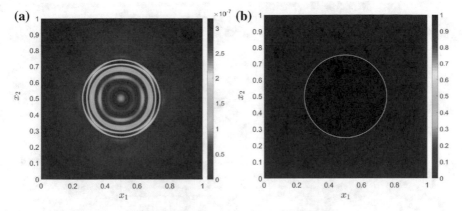

Fig. 4.5 (a) Picture of $|y(u_{det}) - y(u_{exact})|$ at each point $x \in D$, where $y(u_{exact})$ and $y(u_{det})$ are, respectively, the solutions to (4.18), with $u = u_{exact}$ and $u = u_{det}$. (b) Picture of u_{det}

Algorithm for problem (P_1)

Initialization: Take an initial $u^0 \in L^2(D)$, stopping parameters η, δ for (4.5) and (4.40), respectively, and discretization parameters h and ℓ for approximation in the physical and random domains.

Construction: For $k = 0, 1, \cdots$ (until convergence) do:

2.1: Computation of a descent direction $\overline{u}^k = -J_1'\left(u^k\right)$, as given by (4.10), with $u = u^k$. To this end, first solve the direct problem (4.7) and then the adjoint system (4.9).

2.2: Choose an optimal step size parameter λ^k by using expressions (4.11) and (4.12).

2.3: Update the control variable $u^{k+1} = u^k + \lambda^k \overline{u}^k$.

Verification: The stopping criterion is (4.5).

We take $\delta = 10^{-3}$ in (4.40). The enriched level in the adaptive algorithm of Sect. 4.2.2.2 is taken as $\overline{\ell} = 9$ leading to a quadrature level $\ell = 5$. Figure 4.6 displays the rate of convergence for the adaptive, anisotropic, sparse grid algorithm. Gauss-Hermite nodes and weights are used for numerical integration in the random domain. As shown in Table 4.1, this corresponds to 149 nodes in the sparse grid. The mesh for discretization of the physical domain is shown in Fig. 4.3.

The algorithm is initiated with a control $u^0(x) = 0$, $x \in \mathcal{O}$. Figure 4.7b plots the history convergence of the gradient algorithm for the first 100 iterates. Figure 4.8 plots the computed optimal control u_1.

Let us denote by $y(u_{det})$ the solution to the random elliptic problem (4.18) with $u = u_{det}$ the optimal control for the deterministic problem (P_{det}), and by $y(u_1)$ the solution to (4.18), with $u = u_1$ the optimal robust control for (P_1) obtained after 100 iterates of the gradient algorithm. Figure 4.9 presents numerical results for

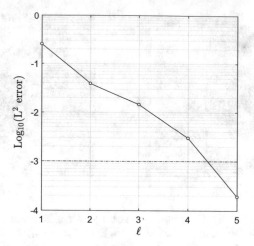

Fig. 4.6 Experiment 1. Rates of convergence of the anisotropic sparse grid algorithm: level of quadrature ℓ versus stopping criterion (4.40). The dashed line represents the prescribed accuracy level $\delta = 10^{-3}$

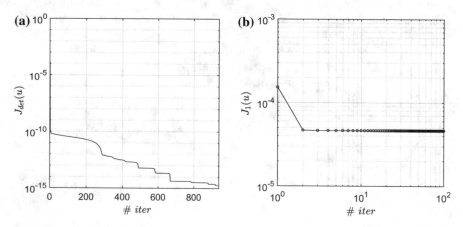

Fig. 4.7 Convergence histories (number of iterations versus values of cost functionals) for the gradient algorithm of Sect. 4.2 corresponding to J_{det} (**a**) and J_1 (**b**)

$\int_\Gamma |y(x, z) - y_d(x)|^2 \rho(z)\, dz$ for the cases: (a) $y = y(u_{det})$ and (b) $y = y(u_1)$. As expected, compared to the optimal deterministic control u_{det}, the optimal control u_1 for problem (P_1), reduces in a significant way the $L_\rho^2(\Gamma)$-distance from $y(x, \cdot)$ to $y_d(x)$.

Figure 4.10 displays the expected value and variance of $y(u_{det})$ (left panel), and of $y(u_1)$ (right panel). It is observed that the the expected values of $y(u_{det})$ and $y(u_1)$ are quite similar. However, the variance of $y(u_1)$ is lower than the one of $y(u_{det})$.

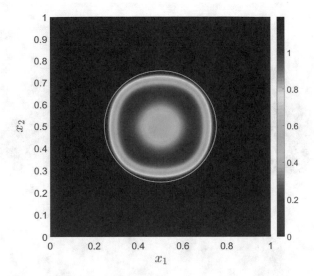

Fig. 4.8 Experiment 1. Control u_1, optimal solution of problem (P_1)

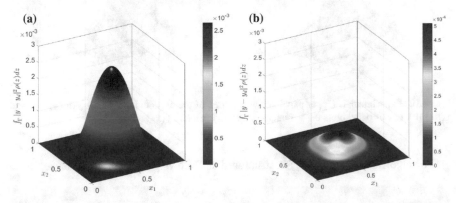

Fig. 4.9 Experiment 1. (a) $\int_\Gamma |y(u_{det}) - y_d|^2 \rho(z)\, dz$, and (b) $\int_\Gamma |y(u_1) - y_d|^2 \rho(z)\, dz$

4.2.3.3 Experiment 2: Problem (P_2)

This section presents numerical simulation results for problem (P_2), with $\gamma = 0$ in (4.13) and for different values of the weighting parameter β. The target function $y_d(x)$ is the same as in the deterministic problem.

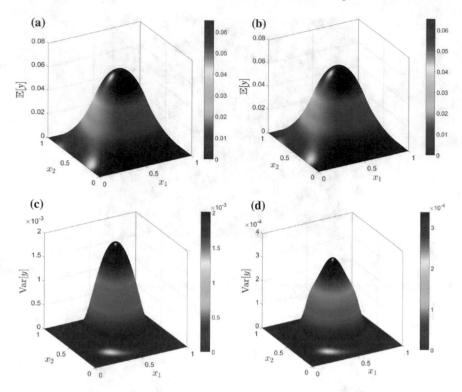

Fig. 4.10 Experiment 1. Upper panel shows pictures of the expected values for $y(u_{det})$ (**a**), and $y(u_1)$ (**b**). Lower panel plots the variances of $y(u_{det})$ (**c**), and $y(u_1)$ (**d**)

The algorithm to solve (P_2) is structured as follows:

Algorithm for problem (P_2)

Initialization: Take an initial $u^0 \in L^2(D)$, stopping parameters η, δ for (4.5) and (4.38), respectively, and discretization parameters h and ℓ for approximation in the physical and random domains.

Construction: For $k = 0, 1, \cdots$ (until convergence) do:

2.1: Computation of a descent direction $\overline{u}^k = -J_2'(u^k)$, as given by (4.14), with $u = u^k$. To this end, first solve the direct problem (4.7) and then the adjoint system (4.15).

2.2: Choose an optimal step size parameter λ^k by using expressions (4.16) and (4.17).

2.3: Update the control variable $u^{k+1} = u^k + \lambda^k \overline{u}^k$.

Verification: The stopping criterion is (4.5).

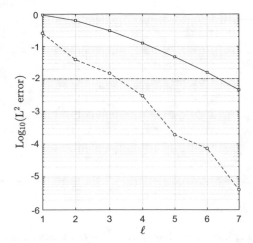

Fig. 4.11 Experiment 2. Rates of convergence of the anisotropic sparse grid algorithm: level of quadrature ℓ versus Log_{10} of the stopping criterion (4.38). The dashed-circle line corresponds to the first term in (4.38), i.e., the one associated to the first statistical moment of y, and the continuous-circle line corresponds to the second term in (4.38), i.e., the one associated to the second moment of y. The horizontal dashed line represents the prescribed accuracy level $\delta = 10^{-2}$

The tolerance in the stopping criterion (4.38) is $\delta = 10^{-2}$. The enriched level in the adaptive algorithm of Sect. 4.2.2.2 is taken as $\bar{\ell} = 8$ leading to a quadrature level $\ell = 7$. This corresponds to 555 Gauss-Hermite nodes in the sparse grid. Figure 4.11 displays the rate of convergence for the adaptive, anisotropic, sparse grid algorithm. The same mesh as in the preceding examples, for discretization of the physical domain D, is used.

The algorithm is initiated with $u^0(x) = 0$, $x \in \mathcal{O}$. The convergence history for the first 100 iterations of the gradient algorithm is shown in Fig. 4.12, for $\beta = 0$ and $\beta = 2$. Figure 4.13 displays the optimal controls for problem (P_2), after 100 iterates of the gradient algorithm, corresponding to different values of the weighting parameter β.

Figure 4.14 shows the expected values and variances of the optimal states for problem (P_2). In the pictures of the left panel, the target function is shown transparently as a reference. As expected, increasing the value of β produces a reduction in variance. The price paid for this variance reduction is that the expected value of y moves further away from the target y_d. This fact is also numerically observed in Table 4.2. First row in Table 4.2 presents values for: $\int_\Gamma \int_D (y - y_d(x))^2 \, dx \rho(z) dz$ (first column), $\int_D (\mathbb{E}[y] - y_d(x))^2 \, dx$ (second column), and $\int_D \text{Var}(y(x)) \, dx$ (third column), which are obtained when the control u_{det} is inserted in the (random) state Eq. (4.18). Second row shows the same quantities but for the optimal control of problem (P_1). The rest of rows show the same quantities for the optimal controls of (P_2) corresponding to $\beta = 0$ (third row), $\beta = 1$ (fourth row), and $\beta = 2$ (fifth row).

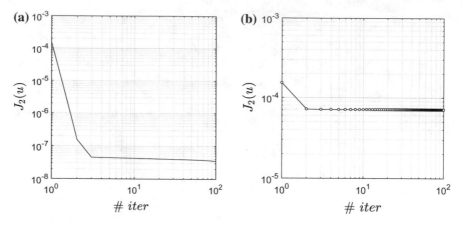

Fig. 4.12 Experiment 2. Convergence histories (number of iterations versus values of cost functionals) for the gradient algorithm of Sect. 4.2 corresponding to J_2 for $\beta = 0$ (**a**) and $\beta = 2$ (**b**)

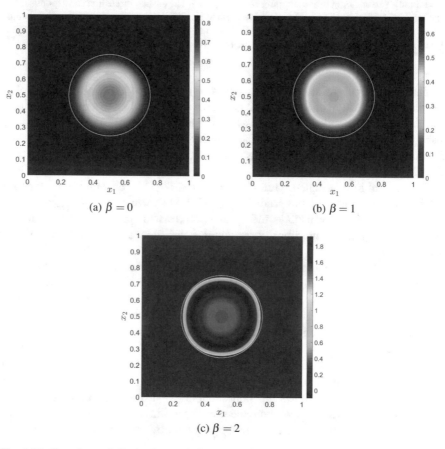

Fig. 4.13 Experiment 2. Optimal controls for $\beta = 0$ (**a**), $\beta = 1$ (**b**), and $\beta = 2$ (**c**)

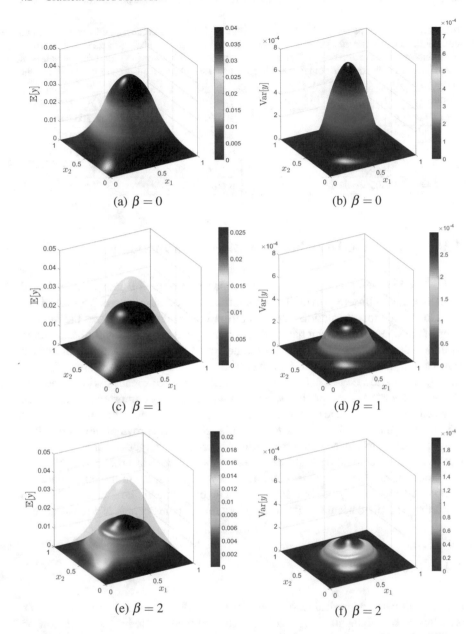

Fig. 4.14 Experiment 2. Expected values and variances of $y(u_2)$, optimal state of (P_2), for different values of β. u_2 denotes the solution to (P_2)

Table 4.2 Summary of results for the cost functional J_1 (second column), and for the first two terms of the cost functional J_2 corresponding to the optimal deterministic control u_{det} (first row), the optimal control u_1 for problem (P_1), and the optimal controls u_2 of (P_2) for different values of β. All integrals in the random domain have been approximated with a quadrature level $\ell = 7$ in the sparse grid

Control	$\int_\Gamma \int_D (y - y_d(x))^2 \, dx \rho(z) dz$	$\int_D (\mathbb{E}[y] - y_d(x))^2 \, dx$	$\int_D \text{Var}(y(x)) \, dx$
u_{det}	4.5422e-04	1.1092e-04	3.4330e-04
u_1	**9.2445e-05**	2.8380e-05	6.4065e-05
$u_2, \beta = 0$	1.3362e-04	**6.9242e-08**	1.3355e-04
$u_2, \beta = 1$	9.2690e-05	2.8197e-05	6.4493e-05
$u_2, \beta = 2$	1.0383e-04	6.6029e-05	**3.7798e-05**

4.3 Benefits and Drawbacks of the Cost Functionals

In this section the costs J_1 and J_2 are compared in terms of their computational efficiency and ability to provide robust solutions. Concerning the capability of the costs to provide robust solutions, the first term of J_1 implicitly penalizes the fluctuations of the state variable which permits to obtain robust solutions without including explicitly the variance. This issue has the advantage of avoiding the tunning of the parameter β, a priori unknown. However, the numerical experiments show that the cost J_1 can be too conservative since it penalizes the differences with respect to the target along the entire stochastic domain which notably compromises the match between the mean state and the target. Conversely, the cost J_2 includes the variance of the state variable as a robustness criterion. This second term provides more flexibility than the cost J_1 to balance the error with respect to the target and the robustness. However, this scalarization requires to solve a set of control problems to find the best trade-off between robustness and accuracy. With respect to the computational complexity, the cost J_2 needs a larger number of stochastic collocation points to fulfill the prescribed accuracy due to the slow rate of convergence of the variance.

4.4 One-Shot Methods

In this section, the numerical resolution of problem (P_1), as given by (4.6) and (4.7), is addressed by solving its associated first-order optimality conditions. From now on in this section, we essentially follow the approach in [16].

As is well-known, these optimality conditions are derived by equating to zero the directional derivatives of the Lagrangian (4.8) with respect to the state, adjoint and control variables. A straightforward computation leads to the coupled system

$$\begin{cases} -\text{div}\,(a\nabla y) = 1_{\mathscr{O}}u, & \text{in } D \times \Gamma \\ y = 0, & \text{on } \partial D \times \Gamma \\ -\text{div}\,(a\nabla p) = y - y_d, & \text{in } D \times \Gamma \\ p = 0, & \text{on } \partial D \times \Gamma \\ \gamma u + \int_{\Gamma} p(x,z)\rho(z)\,dz = 0 & \text{in } \mathscr{O}. \end{cases} \qquad (4.43)$$

Solving in u the last equation and substituting into the first equation gives the reduced optimality system

$$\begin{cases} -\text{div}\,(a\nabla y) = -\frac{1}{\gamma}1_{\mathscr{O}}\int_{\Gamma} p(x,z)\rho(z)\,dz, & \text{in } D \times \Gamma \\ -\text{div}\,(a\nabla p) = y - y_d, & \text{in } D \times \Gamma \\ y = p = 0, & \text{on } \partial D \times \Gamma. \end{cases} \qquad (4.44)$$

Remark 4.8 Well-posedness of the optimality system (4.44) may be easily obtained by applying the Brezzi theory on saddle-point problems [5]. See also [6] for a related problem.

The numerical resolution of the optimality system (4.44) may be addressed by using both a Stochastic Collocation (SC) or a Stochastic Galerkin (SG) finite element method. As it has been illustrated in the preceding section, SC entails the solution of a number of uncoupled deterministic problems. Thus, SC methods are very useful to solve uncoupled problems. However, system (4.44) is coupled via the term $\int_{\Gamma} p(x,z)\rho(z)\,dz$. As a consequence, the advantage associated to decoupled systems, which is associated to the SC method, is lost. It is therefore more convenient to use a SG method, which, typically, requires less degrees of freedom in the random domain than SC for the same accuracy (see [3] and [16, Remarks 1 and 2] for more details).

When applied to problem (4.44), SG finite element methods propose an approximation of the space $L_{\rho}^2(\Gamma) \otimes H_0^1(D)$, where both the state variable $y(x,z)$ and the adjoint state $p(x,z)$ live in, in the form $\mathscr{P}_{\Lambda(\ell,N)}(\Gamma) \otimes V_h(D)$, with

$$\mathscr{P}_{\Lambda(\ell,N)}(\Gamma) = \text{span}\left\{ \psi_q(z) = \prod_{i=1}^{N} \psi_{q_i}(z_i), \quad q = (q_1, \cdots, q_N) \in \Lambda(\ell,N) \right\} \subset L_{\rho}^2(\Gamma)$$

a multivariate polynomial space, and

$$V_h(D) = \text{span}\,\{\phi_i(x), \quad 1 \le i \le M_h\} \subset H_0^1(D),$$

a standard finite element space associated to a given triangulation $\mathscr{T}_h$ of the physical space D.

Here, $\ell \in \mathbb{N}_+$ denotes a level of approximation, and N is the number of terms in the truncated KL expansion of $a(x,z)$, e.g., if $a(x,z)$ is a log-normal random field, then

$$a(x,z) \approx a_N(x,z) = e^{\mu + \sigma \sum_{n=1}^{N} \sqrt{\lambda_n} b_n(x) z_n}, \quad z = (z_1, \cdots, z_N) \in \Gamma. \qquad (4.45)$$

$\Lambda\,(\ell,\,N)$ is a multi-index set, and $\left\{\psi_{q_i}(z_i)\right\}_{q_i=1}^{\infty}$ is a family of orthonormal polynomials in $L^2_{\rho_i}\,(\Gamma_i)$.

Remark 4.9 Similarly to the stochastic collocation method presented in Sect. 4.2.2, when $a(x,z)$ is given by (4.45) it is very convenient to use an anisotropic polynomial space, e.g., the one which will be considered in next chapter. For the sake of simplicity, we do not give more details on the polynomial space $\mathscr{P}_{\Lambda(\ell,N)}\,(\Gamma)$ in this section.

With all these ingredients, a SG formulation of the optimality system (4.44) reads as: find $y_{hq},\,p_{hq}\in\mathscr{P}_{\Lambda(\ell,N)}\,(\Gamma)\otimes V_h\,(D)$ such that

$$
\begin{cases}
\int_\Gamma\int_D a_N\nabla y_{hq}\cdot\nabla v_{hq}\rho\,dxdz+\frac{1}{\gamma}\int_\Gamma\int_D \mathbb{1}_\mathscr{O}\left(\int_\Gamma p_{hq}\rho\,dz\right)v_{hq}\rho\,dxdz=0,\\
\int_\Gamma\int_D a_N\nabla p_{hq}\cdot\nabla w_{hq}\rho\,dxdz-\int_\Gamma\int_D y_{hq}w_{hq}\rho\,dxdz=-\int_\Gamma\int_D y_d w_{hq}\rho\,dxdz,\\
\forall v_{hq},w_{hq}\in\mathscr{P}_{\Lambda(\ell,N)}\,(\Gamma)\otimes V_h\,(D).
\end{cases}
$$

$$(4.46)$$

The matrix formulation of system (4.46) is as follows. Let $M\in\mathbb{R}^{M_h\times M_h}$ be the mass matrix with entries

$$
[M]_{ij}=\int_D \phi_i\phi_j\,dx,\quad 1\le i,j\le M_h,
$$

and let $K_n\in\mathbb{R}^{M_h\times M_h}$, $1\le n\le N$, be a set of stiffness matrices with entries

$$
[K_n]_{ij}=\int_D \sqrt{\lambda_n}b_n\,(x)\,\nabla\phi_i(x)\cdot\nabla\phi_j(x)\,dx,\quad 1\le i,j\le M_h.
$$

Let us denote by $Q=\dim\left(\mathscr{P}_{\Lambda(\ell,N)}\,(\Gamma)\right)$ and consider the set of stochastic matrices $C_n\in\mathbb{R}^{Q\times Q}$, $1\le n\le N$, with entries

$$
[C_n]_{ij}=\int_\Gamma z_n\psi_{\mu(i)}(z)\psi_{\mu(j)}(z)\rho(z)\,dz,\quad 1\le i,j\le Q,
$$

where

$$
\mu:\{1,2,\cdots,Q\}\to\Lambda\,(\ell,N)
$$

is a bijection that assigns a unique integer j to each multi-index $\mu(j)\in\Lambda\,(\ell,N)$.

As noticed in Sect. 2.2, the set

$$
\left\{\psi_{\mu(1)}\phi_1,\cdots,\psi_{\mu(1)}\phi_{M_h},\psi_{\mu(2)}\phi_1,\cdots,\psi_{\mu(2)}\phi_{M_h},\cdots,\psi_{\mu(Q)}\phi_1,\cdots,\psi_{\mu(Q)}\phi_{M_h}\right\}
$$

$$(4.47)$$

is a basis of $\mathscr{P}_{\Lambda(\ell,N)}\,(\Gamma)\otimes V_h\,(D)$. Let us denote by $\mathbf{y}_m=\left(y_{1,m},\cdots,y_{M_h,m}\right)$, $\mathbf{p}_m=\left(p_{1,m},\cdots,p_{M_h,m}\right)\in\mathbb{R}^{M_h}$, with $1\le m\le Q$, the degrees of freedom of y_{hq} and p_{hq}, respectively, i.e.,

$$
y_{hq}\,(x,z)=\sum_{n=1}^{M_h}\sum_{m=1}^{Q}y_{n,m}\phi_n(x)\psi_{\mu(m)}(z),\quad p_{hq}\,(x,z)=\sum_{n=1}^{M_h}\sum_{m=1}^{Q}p_{n,m}\phi_n(x)\psi_{\mu(m)}(z).
$$

Substituting these expressions in (4.46) and replacing v_{hq} and w_{hq} by the elements of the basis (4.47), system (4.46) takes the form

$$\left(\sum_{n=1}^{N} \left[\begin{matrix} C_n & 0_Q \\ 0_Q & C_n \end{matrix} \right] \otimes K_n + \left[\begin{matrix} 0_Q & 0_Q \\ -1_Q & 0_Q \end{matrix} \right] \otimes M + \frac{1}{\gamma} \left[\begin{matrix} 0_{M_h \times Q} & T \\ 0_{M_h \times Q} & 0_{M_h \times Q} \end{matrix} \right] \right) \left(\begin{matrix} \mathbf{y}_1 \\ \cdots \\ \mathbf{y}_Q \\ \mathbf{p}_1 \\ \cdots \\ \mathbf{p}_Q \end{matrix} \right) = \left(\begin{matrix} 0 \\ \cdots \\ 0 \\ \overline{\mathbf{y}}_1^d \\ \cdots \\ \overline{\mathbf{y}}_Q^d \end{matrix} \right)$$

$$(4.48)$$

where $0_Q \in \mathbb{R}^{Q \times Q}$ and $0_{M_h \times Q} \in \mathbb{R}^{(M_h \times Q) \times (M_h \times Q)}$ are zero matrices, $1_Q \in \mathbb{R}^{Q \times Q}$ is the identity matrix,

$$\overline{\mathbf{y}}_j^d = - \left(\int_\Gamma \psi_{\mu(j)} \rho \, dz \int_D y_d \phi_1 \, dx, \cdots, \int_\Gamma \psi_{\mu(j)} \rho \, dz \int_D y_d \phi_{M_h} \, dx \right), \quad 1 \leq j \leq Q,$$

and

$$T = \left[\begin{matrix} T_{11} & \cdots & T_{1Q} \\ \vdots & \ddots & \vdots \\ T_{Q1} & \cdots & T_{QQ} \end{matrix} \right] \in \mathbb{R}^{(M_h \times Q) \times (M_h \times Q)}$$

is a block matrix where the entries of each block $T_{kl} \in \mathbb{R}^{M_h \times M_h}$, $1 \leq k, l \leq Q$, are given by

$$[T_{kl}]_{ij} = \int_\Gamma \psi_{\mu(l)}(z) \rho(z) \int_{\mathcal{O}} \phi_i(x) \phi_j(x) \left(\int_\Gamma \psi_{\mu(k)}(z) \rho(z) \, dz \right) dx dz, \quad 1 \leq i, j \leq M_h.$$

Notation. We recall that given two matrices $A = (a_{ij}) \in \mathbb{R}^{m \times n}$ and $B = (b_{ij}) \in \mathbb{R}^{p \times q}$, $A \otimes B$ denotes the Kronecker product of A and B, which is explicitly given by

$$A \otimes B = \left[\begin{matrix} a_{11}b_{11} & a_{11}b_{12} & \cdots & a_{11}b_{1q} & \cdots\cdots & a_{1n}b_{11} & a_{1n}b_{12} & \cdots & a_{1n}b_{1q} \\ a_{11}b_{21} & a_{11}b_{22} & \cdots & a_{11}b_{2q} & \cdots\cdots & a_{1n}b_{21} & a_{1n}b_{22} & \cdots & a_{1n}b_{2q} \\ \vdots & \vdots & \ddots & \vdots & & \vdots & \vdots & \ddots & \vdots \\ a_{11}b_{p1} & a_{11}b_{p2} & \cdots & a_{11}b_{pq} & \cdots\cdots & a_{1n}b_{p1} & a_{1n}b_{p2} & \cdots & a_{1n}b_{pq} \\ \vdots & \vdots & \vdots & \ddots & \vdots & \vdots & \vdots & \vdots \\ \vdots & \vdots & \vdots & \ddots & \vdots & \vdots & \vdots & \vdots \\ a_{m1}b_{11} & a_{m1}b_{12} & \cdots & a_{m1}b_{1q} & \cdots\cdots & a_{mn}b_{11} & a_{mn}b_{12} & \cdots & a_{mn}b_{mq} \\ a_{m1}b_{21} & a_{m1}b_{22} & \cdots & a_{m1}b_{2q} & \cdots\cdots & a_{mn}b_{21} & a_{mn}b_{22} & \cdots & a_{mn}b_{2q} \\ \vdots & \vdots & \ddots & \vdots & & \vdots & \vdots & \ddots & \vdots \\ a_{m1}b_{p1} & a_{m1}b_{p2} & \cdots & a_{m1}b_{pq} & \cdots\cdots & a_{mn}b_{p1} & a_{mn}b_{p2} & \cdots & a_{mn}b_{pq} \end{matrix} \right].$$

Table 4.3 Monomials up to 2nd degree (in each variable)

		1		
	z_1		z_2	
z_1^2		$z_1 z_2$		z_2^2
	$z_1^2 z_2$		$z_2^2 z_1$	
		$z_1^2 z_2^2$		

The numerical resolution of system (4.48) is typically addressed by using iterative preconditioned Krylov (sub)space solvers [16]. The computational cost, in the random domain, of the SG method presentend in this section is, therefore, given by the product of stochastic degrees of freedom of system (4.48) times the number of iterations required by the iterative solver.

4.5   Notes and Related Software

The sparse grid described in Sect. 4.2.2.1 is usually referred in the literature as to *anisotropic, Smolyak, sparse grid*. Other choices for selecting the number m_ℓ of nodes at the level ℓ and the index set (4.27) are also possible. For a given number of nodes M, the problem of computing the optimal grid, in the sense of minimizing numerical error, is therefore a pertinent one. This problem is being object of an intensive research effort [4, 7]. Next, we briefly describe the main idea underlying Smolyak sparse grid construction.

The idea of Smolyak. The main goal of sparse grids constructions is to keep the same accuracy as full tensor rules but with a reduced number of nodes and functions evaluations. The accuracy of a quadrature rule is quantified in terms of the degree of the polynomials that can be integrated exactly. For instance, to integrate exactly polynomials $p(z_1, z_2)$ of degree less than or equal to 2, in two dimensions, one only need to consider the following monomials:

$$\left\{ 1, z_1, z_1^2, z_2, z_2^2, z_1 z_2 \right\}.$$

Thus, terms like $z_1^2 z_2^2$, that arise in a full tensor product, are unnecessary for the prescribed accuracy. More precisely, the monomials below the line in Table 4.3 are not needed.

In the same vein, the idea of Smolyak was to add low order grids together in such a way that the resulting quadrature rule (as given by (4.29), or equivalently by (4.30)–(4.33)) is expressed as a linear combinations of full tensor grids, but each one of them with a relatively low number of points. It is important to point out that, due to the use of the difference operators (4.28), negative weights may appear in the quadrature rule. All of this is schematically represented in Fig. 4.15.

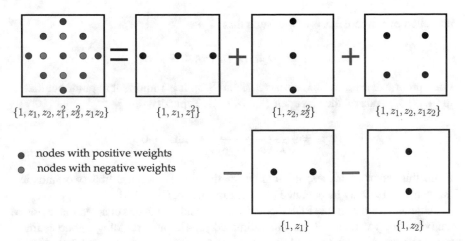

Fig. 4.15 Scheme of the isotropic ($g = 1$) Smolyak quadrature rule (4.30) using Gauss-Hermite nodes in dimension $N = 2$ and level $\ell = 3$

In addition, a Smolyak's quadrature rule that employs nested points requires fewer function evaluations than the corresponding formula with non-nested points. Anyway, the major computational savings come from the fact that most of the points in the full tensor grid are removed. This improvement is greater as the dimension and/or the level increase.

Smoothness and convergence. Convergence results of the collocation method presented in this chapter depend on the regularity of the solution $y(x, z)$ of system (4.18), w.r.t. the random parameter $z \in \Gamma$. Indeed, with the same notations as in Sect. 4.1, let us consider the compact set

$$\Gamma_n^* = \prod_{j=1, j \neq n}^{N} \Gamma_j,$$

and let us denote by z_n^* an element in Γ_n^*. For continuous diffusion coefficients,[2] i.e., $a \in C^0 (\Gamma; L^\infty (D))$, then the solution $y(x, z)$ to (4.18), considered as a function of the variable $z_n \in \Gamma_n$ with values in the space of continuous functions from Γ_n^* to $H_0^1 (D)$, i.e.,

$$y = y\left(x, \cdot, z_n^*\right) : \Gamma_n \to C^0 \left(\Gamma_n^*; H_0^1 (D)\right),$$

admits an analytical extension $y\left(x, s, z_n^*\right), s \in \mathbb{C}$, in the region of the complex plane

$$\sum \left(\Gamma_n; \tau_n\right) = \{s \in \mathbb{C} : \text{dist}\left(s, \Gamma_n\right) \leq \tau_n\}, \tag{4.49}$$

---

[2]For instance, when $a(x, z)$ is expressed as a truncated KL expansion of a log-normal randon field (see [2, Example 3]).

and, moreover, there exists $c > 0$, such that

$$\| y\,(s) \|_{C^0(\Gamma_n^*; H_0^1(D))} \leq c.$$

For a proof, we refer to [2, Lemmas 3.1. and 3.2]. It is important to point out that τ_n in (4.49) depends on the eigenpair $\{\lambda_n, b_n\,(x)\}$. Typically,

$$\tau_n = \frac{\alpha}{\sqrt{\lambda_n}\|b_n\|_{L^\infty(D)}}, \quad \text{with } \alpha > 0.$$

From this, error estimates, in suitable spatial norms, for the first two statistical moments (4.19) may be obtained (see [2, Lemmas 4.7 and 4.8]).

Although for the simple PDE model (4.18) considered in this chapter, the region of analyticity of y w.r.t. $z \in \Gamma$ may be estimated [2, 4], thus providing a good estimate for the vector g, as given by (4.25), in many other cases it might not be possible to sharply compute g using a priori information. An easily implementable method for selecting g using a posteriori information is presented in [12].

The methods presented in this chapter have a nice computational performance when the number of random inputs is not so large. However, a common critical computational challenge which faces most of stochastic collocation methods is *high dimensionality*. Indeed, when the number N of input random variables becomes large, the number of stochastic degrees of freedom required for a prescribed accuracy level might grow exponentially when isotropic (or even anisotropic) sparse grids are used. This phenomenon is known as the *curse of dimensionality*. A number of methods to overcome, or at least to alleviate, this undesirable performance have been proposed during the last years (see [1, 7, 8] and the references therein).

For sparse grids constructions and quadrature rules used in all experiments in this chapter we have used the Sparse grids Matlab kit [3] (http://csqi.epfl.ch). For finite element analysis we have used the Matlab vectorized codes provided by [14]. We are indebted to the authors for letting us the use of these two packages.

In the line of Sect. 4.3, for a comparison study of different functionals for random PDE optimization problems we refer to [10].

Apart from the codes accompanying this book, the interested reader may find additional open source codes for interpolation and quadrature rules on sparse grids in the links provided below:

- Sparse grids Matlab kit [3] provided by Scientific Computing and Uncertainty Quantification (CSQI) group. http://csqi.epfl.ch.
- The Dakota software project developed by Sandia National Laboratories. Provides state-of-the-art research and robust, usable software for optimization and uncertainty quantification (UQ). https://dakota.sandia.gov/
- The Sparse Grid Interpolation Toolbox is a Matlab toolbox for recovering (approximating) expensive, possibly high-dimensional multivariate functions. It was developed by Andreas Klimke at the Institute of Applied Analysis and Numerical

Simulation, High Performance Scientific Computing lab, Universität Stuttgart. http://www.ians.uni-stuttgart.de/spinterp/
- Computer C++ code in support of paper [16]. Very well-suited for Sect. 4.4. https://www.repository.cam.ac.uk/handle/1810/240740
- The Toolkit for Adaptive Stochastic Modeling and Non-Intrusive Approximation is a collection of robust libraries for high dimensional integration and interpolation as well as parameter calibration. The project is sponsored by Oak Ridge National Laboratory Directed Research and Development as well as the Department of Energy Office for Advanced Scientific Computing Research. http://tasmanian.ornl.gov/
- Several useful codes supporting different programming language are provided by Professor John Burkardt from Department of Scientific Computing at Florida State University (FSU). https://people.sc.fsu.edu/ jburkardt/

References

1. Alexanderian, A., Petra, N., Stadler, G., Ghattas, O.: Mean-variance risk-averse optimal control of systems governed by PDEs with random parameter fields using quadratic approximations. SIAM/ASA J. Uncertain. Quantif. **5**(1), 1166–1192 (2017)
2. Babuška, I., Nobile, F., Tempone, R.: A Stochastic collocation method for elliptic partial differential equations with random input data. SIAM Rev. **52**(2), 317–355 (2010)
3. Bäck, J., Nobile, F., Tamellini, L., Tempone, R.: Stochastic spectral Galerkin and collocation methods for PDEs with random coefficients: a numerical comparison. In: Spectral and High Order Methods for Partial Differential Equations. Lecture Notes in Computational Science and Engineering, vol. 4362, p. 76. Springer, Heidelberg (2011)
4. Bäck, J., Tempone, R., Nobile, F., Tamellini, L.: On the optimal polynomial approximation of stochastic PDEs by Galerkin and collocation methods. Math. Models Methods Appl. Sci. **22**(9), 1250023 (2012)
5. Brezzi, F.: On the existence, uniqueness and approximation of saddle-point problems arising from lagrangian multipliers. RAIRO Modél. Math. Anal. Numér. **8**(2), 129–151 (1974)
6. Chen, P., Quarteroni, A., Rozza, G.: Stochastic optimal Robin boundary control problems of advection-dominated elliptic equations. SIAM J. Numer. Anal. **51**(5), 2700–2722 (2013)
7. Chen, P., Quarteroni, A.: A new algorithm for high-dimensional uncertainty quantification based on dimension-adaptive sparse grid approximation and reduced basis methods. J. Comput. Phys. **298**, 176–193 (2015)
8. Chen, P., Quarteroni, A., Rozza, G.: Comparison between reduced basis and stochastic collocation methods for elliptic problems. J. Sci. Comput. **59**(1), 187–216 (2014)
9. Hinze, M., Pinnau, R., Ulbrich, M., Ulbrich, S.: Optimization with PDE constraints. In: Mathematical Modelling: Theory and Applications, vol. 23. Springer, Berlin (2009)
10. Lee, HCh., Gunzburger, M.D.: Comparison of approaches for random PDE optimization problems based on different matching functionals. Comput. Math. Appl. **73**(8), 1657–1672 (2017)
11. Lord, G.L., Powell, C.E., Shardlow, T.: An Introduction to Computational Stochastic PDEs. Cambridge University Press (2014)
12. Nobile, F., Tempone, R., Webster, C.G.: An anisotropic sparse grid stochastic collocation method for partial differential equations with random input data. SIAM J. Numer. Anal. **46**(5), 2411–2442 (2008)
13. Novak, E., Ritter, K.: High-dimensional integration of smooth functions over cubes. Numer. Math. **75**(1), 79–97 (1996)
14. Rahman, T., Valdman, J.: Fast MATLAB assembly of FEM matrices in 2D and 3D: nodal elements. Appl. Math. Comput. **219**, 7151–7158 (2013)

15. Rao, M.M., Swift, R.J.: Probability Theory with Applications, 2nd edn. Springer, New York (2006)
16. Rosseel, E., Wells, G.N.: Optimal control with stochastic PDE constraints and uncertain controls. Comput. Methods Appl. Mech. Eng. **213**(216), 152–167 (2012)
17. Smith, R.C.: Uncertainty Quantification. Theory, Implementation, and Applications. Computational Science & Engineering, vol. 12. SIAM, Philadelphia (2014)
18. Smolyak, S.: Quadrature and interpolation formulas for tensor product of certain classes of functions. Dokl. Akad. Nauk SSSR **4**, 240–243 (1963)
19. Tröltzsch, F.: Optimal Control of Partial Differential Equations: Theory, Methods and Applications. Graduate Studies in Mathematics, vol. 112. AMS. Providence, Rhode Island (2010)
20. Trefethen, L.N.: Is Gauss quadrature better than Clenshaw Curtis ? SIAM Rev. **50**, 67–87 (2008)

Chapter 5
Numerical Resolution of Risk Averse Optimal Control Problems

> *Far better an approximate answer to the right question, which is often vague, than an exact answer to the wrong question, which can always be made precise.*
>
> John Wilder Tukey.
> The Future of Data Analysis, 1962.

In this chapter, an adaptive, gradient-based, minimization algorithm is proposed for the numerical resolution of the risk averse optimal control problem (P_ε), as defined in Sect. 3.2.2. The numerical approximation of the associated statistics combines an adaptive, anisotropic, non-intrusive, Stochastic Galerkin approach for the numerical resolution of the underlying state and adjoint equations with a standard Monte-Carlo (MC) sampling method for numerical integration in the random domain.

5.1 An Adaptive, Gradient-Based, Minimization Algorithm

Let $y_d \in L^2(D)$ be a given target and $\varepsilon > 0$ a threshold parameter. As indicated in the proof of Theorem 3.5 and taking into account Sect. 4.1, the risk-averse optimal control problem (P_ε), as formulated in Sect. 3.2.2, may be written in the form

$$
\begin{cases}
\text{Minimize in } u : J_\varepsilon(u) = \int_\Gamma H\left(\|y(z) - y_d\|^2_{L^2(D)} - \varepsilon\right) \rho(z)\, dz \\
\text{subject to} \\
\qquad y = y(u) \quad \text{solves (4.18)}, \quad u \in \mathscr{U}_{ad},
\end{cases}
\tag{5.1}
$$

where H is the Heaviside function and, for simplicity, $\mathscr{U}_{ad} = L^2(D)$.

© The Author(s), under exclusive license to Springer Nature Switzerland AG 2018
J. Martínez-Frutos and F. Periago Esparza, *Optimal Control of PDEs under Uncertainty*,
SpringerBriefs in Mathematics, https://doi.org/10.1007/978-3-319-98210-6_5

Since the Heaviside step function $H(s)$ is not differentiable at $s = 0$, the smooth approximation

$$H_\alpha(s) = \left(1 + e^{-\frac{2s}{\alpha}}\right)^{-1}, \quad \text{with } 0 < \alpha < 1$$

is considered. Thus, the cost functional $J_\varepsilon(u)$ is replaced by

$$J_\varepsilon^\alpha(u) = \int_\Gamma \left[1 + e^{-\frac{2}{\alpha}\left(\|y(z)-y_d\|^2_{L^2(D)}-\varepsilon\right)}\right]^{-1} \rho(z)\,dz \tag{5.2}$$

Eventually, problem (5.1) is approximated by

$$\begin{cases} \text{Minimize in } u: J_\varepsilon^\alpha(u) = \int_\Gamma \left[1 + e^{-\frac{2}{\alpha}\left(\|y(z)-y_d\|^2_{L^2(D)}-\varepsilon\right)}\right]^{-1} \rho(z)\,dz \\ \text{subject to} \\ \qquad\qquad y = y(u) \quad \text{solves (4.18)}, \quad u \in \mathcal{U}_{ad}. \end{cases} \tag{5.3}$$

In addition to the non-differentiability issue mentioned above, another difficulty which arises when trying to implement, in risk averse problems, the gradient algorithm proposed at the beginning of Sect. 4.2 is the following one: denoting by u^k the control at iterate k and by $y_k(z)$ the solution to (4.18) for the control u^k, it is not surprising that the PDF of $\|y_k(\cdot) - y_d\|^2_{L^2(D)}$ (typically at the first iterate) be concentrated in the region where (the approximation of) the Heaviside function is constant and equal to one (i.e., $\|y_k(z) - y_d\|^2_{L^2(D)} > \varepsilon$ for all $z \in \Gamma$). In such a case, $\left(J_\varepsilon^\alpha\right)'(u^k) = 0$ and hence no descent direction is available.

A possibility to overcome this difficulty is to adapt, at each iterate k, the parameters ε and α, which appear in (5.2), to the current location of the PDF of $\|y_k(\cdot) - y_d\|^2_{L^2(D)}$. More precisely, at iterate k, the parameter ε_k may be taken as the mean value of $\|y_k(\cdot) - y_d\|^2_{L^2(D)}$, and α_k proportional to the standard deviation of the same quantity $\|y_k(\cdot) - y_d\|^2_{L^2(D)}$. Both parameters are then introduced in (5.2). Thus, in order to reduce the probability of $\|y_k(z) - y_d\|^2_{L^2(D)}$ to exceeding ε_k, the optimization algorithm (for ε_k and α_k kept fixed) moves the PDF of the current $\|y(\cdot) - y_d\|^2_{L^2(D)}$ towards the left of ε_k. This way, the new updated ε_{k+1} would be lower than ε_k. At the same time, since $\|y(z) - y_d\|^2_{L^2(D)}$ is non-negative, the dispersion of the new $\|y_{k+1}(\cdot) - y_d\|^2_{L^2(D)}$ would also be lower than the previous one. See Fig. 5.1 for a graphical illustration of this idea.

This observation motivates the following minimization algorithm for risk averse optimal control problems:

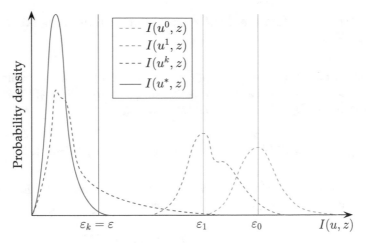

Fig. 5.1 Evolution of the probability density function of $I\left(u^k, z\right) := \|y_k(z) - y_d\|^2_{L^2(D)}$ along the minimization algorithm. Here u^* denotes the optimal risk averse control

Adaptive, gradient-based, optimization algorithm
Initialization: Take an initial $u^0 \in \mathcal{U}$.
Construction: For $k = 0, 1, \ldots$ (while $\varepsilon_k > \varepsilon$) do:
  2.1: Update parameters ε and α as

$$\begin{cases} \varepsilon_k = \text{ mean value of } \|y_k\left(\cdot\right) - y_d\|^2_{L^2(D)} \\ \alpha_k = 0.01 \times \text{ standard deviation of } \|y_k\left(\cdot\right) - y_d\|^2_{L^2(D)}. \end{cases}$$

  2.2: Take as descent direction $\overline{u}^k = -\left(J^{\alpha_k}_{\varepsilon_k}\right)'\left(u^k\right)$.
  2.3: Choose a step size λ^k such that $J^{\alpha_k}_{\varepsilon_k}\left(u^k + \lambda^k\overline{u}^k\right) < J^{\alpha_k}_{\varepsilon_k}\left(u^k\right)$.
  2.4: Set $u^{k+1} = u^k + \lambda^k\overline{u}^k$.
  2.5: Go to 2.1.
When $\varepsilon_k \leq \varepsilon$, both ε_k and α_k are kept fixed ($\varepsilon_k = \varepsilon$ and α_k equal to its current value) and the algorithm continues with steps 2.2 to 2.4.
Verification: The stopping criterion is the first $k \in \mathbb{N}$ for which

$$\frac{|J^{\alpha}_{\varepsilon}\left(u^k\right) - J^{\alpha}_{\varepsilon}\left(u^{k-1}\right)|}{J^{\alpha}_{\varepsilon}\left(u^k\right)} \leq \eta, \tag{5.4}$$

where η is a prescribed tolerance.
If condition $\varepsilon_k \leq \varepsilon$ is not satisfied after a given, large enough, number of iterates, then the algorithm stops. This is an indication that the prescribed ε is too small to be satisfied with controls $u \in \mathcal{U}_{ad}$.

5.2 Computing Gradients of Functionals Measuring Risk Aversion

The gradient of $J_\varepsilon^\alpha (u)$, which is required in step 2.2. of the above minimization algorithm, may be obtained by following the same lines as in Sect. 4.2.1. Precisely, for $\hat{y}, \hat{p} \in L_\rho^2 \left(\Gamma; H_0^1 (D)\right)$, and $\hat{u} \in L^2 (D)$, the Lagrangian

$$\mathscr{L} \left(\hat{y}, \hat{u}, \hat{p}\right) = J_\varepsilon^\alpha \left(\hat{y}, \hat{u}\right) - \int_\Gamma \int_D \left[a\nabla\hat{y} \cdot \nabla\hat{p} - 1_\mathcal{O}\hat{u}\hat{p}\right] dx\rho(z)dz, \qquad (5.5)$$

is considered. Let us denote by (y, u, p) a stationary point of $\mathscr{L}$. Equating the derivative of $\mathscr{L}$ w.r.t. y to zero gives the adjoint system

$$\begin{cases} -\mathrm{div}\,(a\nabla p) = C\,(z)\,(y - y_d), & \text{in } D \times \Gamma \\ p = 0, & \text{on } \partial D \times \Gamma, \end{cases} \qquad (5.6)$$

with

$$C\,(z) = \frac{4}{\alpha}e^{-\frac{2}{\alpha}\left(\|y(z)-y_d\|_{L^2(D)}^2 -\varepsilon\right)} \left[1 + e^{-\frac{2}{\alpha}\left(\|y(z)-y_d\|_{L^2(D)}^2 -\varepsilon\right)}\right]^{-2}. \qquad (5.7)$$

The gradient of J_ε^α is obtained by differentiating the Lagrangian w.r.t. the control u, which leads to

$$\left(J_\varepsilon^\alpha\right)' (u) (x) = \int_\Gamma p\,(x, z)\,\rho(z)dz, \quad x \in \mathcal{O}, \qquad (5.8)$$

where p solves (5.6).

5.3 Numerical Approximation of Quantities of Interest in Risk Averse Optimal Control Problems

Although the random variable $C\,(z)$, which appears in (5.7), is *theoretically* smooth, for α small, numerically it is not. The same problem appears when evaluating the cost functional $J_\varepsilon^\alpha (u)$ and its sensitivity at each iteration of the descent algorithm. As a consequence, stochastic collocation methods are not the best choice to compute the involved integrals in the random domain Γ because if a few collocation nodes are taken in the random discontinuities, then the corresponding numerical approximations could be very inaccurate. It is then more convenient to use a MC sampling strategy. However, the use of a direct MC method requires the numerical resolution of (4.18) and (5.6) at a large number of random sampling points $z^k \in \Gamma$ and at each iteration of the descent method. This makes a direct MC method unaffordable from a computational point of view. For this reason, the MC method will be combined with a Stochastic Galerkin (SG) (also called Polynomial Chaos (PC)) method for uncertainty propagation. More precisely, $y(x, z)$, which is smooth w.r.t. $z \in \Gamma$, will be approximated with a SG method, which is computationally cheaper than MC method. The same method will be applied to approximate the adjoint state $p(x, z)$, solution to (5.6). Then, the MC method will be applied to these two approximations in order to compute the integrals in the random domain, which appear in the cost functional $J_\varepsilon^\alpha (u)$ and in its sensitivity. Next subsection provides details on all these issues.

5.3.1 An Anisotropic, Non-intrusive, Stochastic Galerkin Method

As in the preceding chapter, having in mind that in most of applications the uncertain coefficient $a (x, z)$ is represented as a truncated KL expansion of a log-normal random field, in what follows it is assumed that the different stochastic directions do not have the same influence in the solution $y (x, z)$ to (4.18). As a consequence, it is natural to use an anisotropic SG method to approximate numerically $y (x, z)$ and its associated adjoint state $p (x, z)$, solution to (5.6). Following [2], this subsection presents the construction of such anisotropic polynomial space. We follow an algorithmic description of the method and refer the reader to [2] for error estimates.

With the same notations as in the preceding chapters, let $\left\{ \psi_{p_n} (z_n) \right\}_{p_n=0}^{\infty}, 1 \le n \le N$, be an orthonormal basis of $L_{\rho_n}^2 (\Gamma_n)$ composed of a suitable class of orthonormal polynomials.[1] Since $L_\rho^2 (\Gamma) = \bigotimes_{n=1}^{N} L_{\rho_n}^2 (\Gamma_n)$, a multivariate orthonormal polynomial basis of $L_\rho^2 (\Gamma)$ is constructed as

$$\left\{ \psi_p (z) = \prod_{n=1}^{N} \psi_{p_n} (z_n) \right\}_{p=(p_1,\dots,p_N) \in \mathbb{N}^N}.$$

The construction of an anisotropic multivariate polynomial space, which should be able to keep accuracy and reduce computational cost, is based on a suitable selection of the degrees p_n of the orthonormal polynomials in each stochastic direction. To this end, and similarly to the anisotropic stochastic collocation method described in the preceding chapter, a positive integer $\ell \in \mathbb{N}_+$ indicating the level of approximation in the random space $L_\rho^2 (\Gamma)$, the vector of weights $g = (g_1, \dots, g_N) \in \mathbb{R}_+^N$ for the different stochastic directions (which for the specific model example considered in this text is given by (4.25)) and the multi-index set

$$\Lambda_g (\ell, N) = \left\{ p = (p_1, \dots, p_N) \in \mathbb{N}^N \; : \; \sum_{n=1}^{N} p_n g_n \le \ell \underline{g} \right\}, \qquad (5.9)$$

---

[1] In the numerical experiments of Sect. 5.4, we take $\rho_n (z_n) = \phi (z_n)$, where $\phi (z_n)$ is given by (2.30).

with $\underline{g} = \min_{1 \le n \le N} g_n$, are considered. Finally, the anisotropic approximation multivariate polynomial space is defined as

$$\mathscr{P}_{\Lambda_g(\ell,N)}(\Gamma) = \text{span} \left\{ \psi_p(z), \quad p \in \Lambda_g(\ell, N) \right\}.$$

This choice of $\Lambda_g(\ell, N)$ is referred in the literature [2] as to Anisotropic Total Degree polynomial space (ATD).

Thus, an approximate solution $y_\ell(x, z) \in \mathscr{P}_{\Lambda_g(\ell,N)}(\Gamma) \otimes H_0^1(D)$ of (4.18) is expressed in the form

$$y(x, z) \approx y_\ell(x, z) = \sum_{p \in \Lambda_g(\ell,N)} \hat{y}_p(x) \psi_p(z), \quad \hat{y}_p \in H_0^1(D), \tag{5.10}$$

where due to the orthonormality of $\left\{ \psi_p(z) \right\}_{p \in \Lambda_g(\ell,N)}$,

$$\hat{y}_p(x) = \int_\Gamma y_\ell(x, z) \psi_p(z) \rho(z) \, dz \approx \int_\Gamma y(x, z) \psi_p(z) \rho(z) \, dz. \tag{5.11}$$

This latter integral is numerically approximated using the anisotropic sparse grid collocation method described in the preceding chapter. Eventually, we get a fully discretized solution

$$y_\ell^h(x, z) = \sum_{p \in \Lambda_g(\ell,N)} \hat{y}_p^h(x) \psi_p(z) \in \mathscr{P}_{\Lambda_g(\ell,N)}(\Gamma) \otimes V_h(D)$$

with $V_h(D)$ a standard finite element space approximating $H_0^1(D)$, and

$$\hat{y}_p^h(x) = \int_\Gamma y^h(x, z) \psi_p(z) \rho(z) \, dz \approx \mathscr{A}_g(\ell + 1, N) \, y^h(x) \tag{5.12}$$

a quadrature rule as in (4.35). Here, $y^h(\cdot, z) \in V_h(D)$ denotes the finite element approximation of $y(\cdot, z)$ at the point $z \in \Gamma$.

Notice that the multi-index set used in the quadrature rule (5.12) is $Y_g(\ell + 1, N)$, which is suitable to integrate, by using Gauss-Hermite nodes and weights, polynomials in $\mathscr{P}_{\Lambda_g(\ell,N)}(\Gamma)$.

Remark 5.1 We emphasize that to approximate numerically (5.12), the solution y to (4.18) has to be computed at a number of collocation nodes $z^k \in \Gamma$. This computational task may, therefore, be carried out in parallel computers. Due to the regularity of y w.r.t. z, the number of sampling points z^k required by the stochastic collocation method is lower than the one needed by the Monte Carlo method.

5.3.2 *Adaptive Algorithm to Select the Level of Approximation*

The level of approximation ℓ is adaptively chosen in a similar fashion as in Sect. 4.2.2.2. For the sake of clarity and completeness, it is briefly described next.

Adaptive algorithm to select the level ℓ of approximation

1. Computation of an *enriched* solution $y_{\bar{\ell}}^{h}(x, z)$ of (4.18) by following the method described above for approximation in the randon domain Γ and by using finite elements for discretization in the physical domain D. This enriched solution (which plays the role of *exact solution*) is then used to obtain an approximation $J_{\varepsilon,\bar{\ell}}^{\alpha,h}(u)$ of $J_{\varepsilon}^{\alpha}(u)$ by using MC method, where random samplings z^{k} are applied to $y_{\bar{\ell}}^{h}(x, z)$. As in the preceding chapter, the reference control $u = u_{det}$ is taken as the solution of the associated deterministic optimal control problem (4.42).

2. The level ℓ is linearly increased (from $\ell = 1$ to $\ell = \ell_{opt} < \bar{\ell}$) until the stopping criterion

$$\frac{|J_{\varepsilon,\ell}^{\alpha,h}(u) - J_{\varepsilon,\bar{\ell}}^{\alpha,h}(u)|}{J_{\varepsilon,\bar{\ell}}^{\alpha,h}(u)} \le \delta, \tag{5.13}$$

where $0 < \delta \ll 1$ is a prescribed tolerance, is satisfied.

In the case that the stopping criterion (5.13) does not hold for any $\ell < \bar{\ell}$, then a larger $\bar{\ell}$ is taken to ensure that the enriched solution $y_{\bar{\ell}}^{h}(x, z)$ is good enough. The whole process is then repeated taking the new value of $\bar{\ell}$ as reference.

The use of a Monte Carlo sampling method to approximate the integrals in the random domain Γ, which appear in the cost functional (5.2) and in its sensitivity (5.8), introduces an additional approximation error. Therefore, the number of MC sampling points used must be properly chosen.

5.3.3 *Choosing Monte Carlo Samples for Numerical Integration*

Let $F : \Gamma \to \mathbb{R}$ be a multivariate function. When applied to the numerical approximation of

$$I := \mathbb{E}[F(z)] = \int_{\Gamma} F(z) \rho(z) \, dz, \tag{5.14}$$

the classical Monte Carlo method amounts to drawing M random points $z^k \in \Gamma$, which are distributed according to the PDF $\rho(z)$, and take

$$\frac{1}{M} \sum_{k=1}^{M} F\left(z^k\right) \tag{5.15}$$

as an approximation of (5.14).

This approach is supported by the Strong Law of Large Numbers. Indeed, let $\{\xi_j : \Omega \to \Gamma\}_{j \geq 1}$ be a sequence of i.i.d. random variables and let $\rho = \rho(z)$ its associated joint probability density function (see Sect. 4.1 for details). Consider the sample average

$$\overline{I}_M := \frac{1}{M} \sum_{j=1}^{M} F\left(\xi_j\right). \tag{5.16}$$

Then, by the Strong Law of Large Numbers [7, Chap. 20],

$$\lim_{M \to \infty} \overline{I}_M = I \quad a.s. \quad \text{and in } L_\rho^2(\Gamma).$$

Now, given small constants $TOL_{MC} > 0$ and $\delta_{MC} > 0$, this section addresses the problem of choosing $M = M(TOL_{MC}, \delta_{MC})$ large enough so that the error probability satisfies

$$\mathbb{P}\left(|\overline{I}_M - I| > TOL_{MC}\right) \leq \delta_{MC}. \tag{5.17}$$

Denoting by σ^2 the variance of $F\left(\xi_j\right)$, the Central Limit Theorem [7, Chap 21] states that

$$\frac{\sqrt{M}\left(\overline{I}_M - I\right)}{\sigma} \rightharpoonup \nu \quad \text{in distribution}, \tag{5.18}$$

where $\nu \sim \mathcal{N}(0, 1)$ follows a standard Gaussian distribution, i.e.,

$$\lim_{M \to \infty} \mathbb{E}\left[g\left(\frac{\sqrt{M}\left(\overline{I}_M - I\right)}{\sigma}\right)\right] = \mathbb{E}\left[g\left(\nu\right)\right],$$

for all bounded and continuous functions g.

As a consequence (see e.g. [7, Th. 18.4]), for M large enough,

$$\begin{aligned}
\mathbb{P}\left(|\overline{I}_M - I| > TOL_{MC}\right) &= \mathbb{P}\left(\left|\frac{\sqrt{M}(\overline{I}_M - I)}{\sigma}\right| > \frac{\sqrt{M}TOL_{MC}}{\sigma}\right) \\
&= 1 - \mathbb{P}\left(\left|\frac{\sqrt{M}(\overline{I}_M - I)}{\sigma}\right| \leq \frac{\sqrt{M}TOL_{MC}}{\sigma}\right) \\
&\approx 1 - \mathbb{P}\left(|\nu| \leq \frac{\sqrt{M}TOL_{MC}}{\sigma}\right) \\
&= 2\left(1 - \Phi\left(\frac{\sqrt{M}TOL_{MC}}{\sigma}\right)\right).
\end{aligned}$$

All these arguments motivate the following sample variance based algorithm [3, 6] for selecting M such that (5.17) holds:

Sample variance based algorithm to select M

Input: $TOL_{MC} > 0, \delta_{MC} > 0$, initial number of samples M_0 and the cumulative distribution function of the standard Gaussian variable $\Phi(x)$.

Output: $\overline{I}_M$.

Set $n = 0$, generate M_n samples z^k and compute the sample variance

$$\overline{\sigma}_{M_n}^2 = \frac{1}{M_n - 1} \sum_{k=1}^{M_n} \left(F\left(z^k\right) - \overline{I}_{M_n}\right)^2. \qquad (5.19)$$

while $2\left(1 - \Phi\left(\sqrt{M_n}TOL_{MC}/\overline{\sigma}_{M_n}\right)\right) > \delta_{MC}$ **do**
  Set $n = n + 1$ and $M_n = 2M_{n-1}$.
  Generate a batch of M_n i.i.d. samples z^k.
  Compute the sample variance $\overline{\sigma}_{M_n}^2$ as given by (5.19).
end while
Set $M = M_n$, generate samples z^k, $1 \leq k \leq M$ and compute the output $\overline{I}_M$ as given by (5.15).

5.4 Numerical Experiments

From now on in this Section, it is assumed that the physical domain $D = (0, 1)^2$ is the unit square and the control region is a circle centred at the point $(0.5, 0.5)$ and of radius $r = 0.25$. The uncertain parameter $a(x, z)$ is the truncated KL expansion considered in (2.29), with $N = 5$, which lets capture 89% of the energy field. The correlation lengths are $L_1 = L_2 = 0.5$, and the mean and variance of the random field are constant and equal to 1. The random domain is $\Gamma = [-3, 3]^5$, and

$$\rho(z) = \prod_{n=1}^{5} \phi(z_n), \quad z = (z_1, \ldots, z_5) \in \Gamma, \qquad (5.20)$$

where $\phi(z_n)$ is the truncated Gaussian (2.30). The target function y_d is the same as in the preceding chapter (see Figure 4.4a).

The threshold parameter ε is taken as $\varepsilon = 10^{-4}$. The vector of weights g, which is needed to construct the multi-index set (5.9) is obtained from (4.25) and is given by

$$g = (0.3894, 0.5478, 0.5478, 0.7708, 0.8219). \qquad (5.21)$$

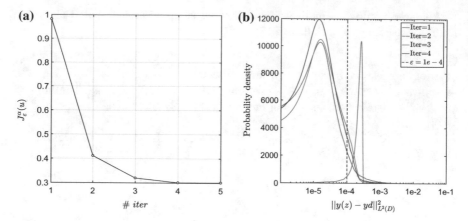

Fig. 5.2 (**a**) Convergence history of the algorithm, and (**b**) evolution of the PDF of the random variable $\|y(z) - y_d\|_{L^2(D)}$

The enriched level used in the adaptive algorithm of Sect. 5.3.2 is taken as $\bar{\ell} = 7$, which, for a tolerance $\delta = 10^{-3}$ in (5.13), leads to $\ell = 5$.

The number M of random Monte Carlo sampling points, which are used for numerical integration of the cost functional and its sensitivity, is taken as $M = 10^5$. For the case of the cost functional (5.2) and being $y = y(u_{det})$ the state associated to u_{det} (solution to problem (4.42)), the stopping criterion in the sample variance algorithm of Sect. 5.3.3 is satisfied with $TOL_{MC} = 0.005$ and $\delta_{MC} = 10^{-3}$.

The step-size parameter λ^k is chosen at each iterate in such a way that it ensures the decrease of the cost function. In the numerical experiments that follow the selection of λ^k is done using a straightforward strategy: in a first stage, an acceptable step length (acceptable meaning that the cost functional decreases) is found after a finite number of trials. Once an acceptable step length is achieved, a second stage is performed to search a value that improves this step length. In this new search, the acceptable step length is increased by multiplying it by a scale factor while the cost functional is reduced w.r.t. the previous iteration. The reader is referred to the computer codes that accompany the text for the specific test examples here included and to [12, Chapter 3] for many other possibilities to select λ^k.

The algorithm is initiated with a constant control $u^0(x) = 0.1$, $x \in \mathcal{O}$. Since for this initial control, $0 < J_\varepsilon(u^0) < 1$ (as it may be observed in Fig. 5.2b), the parameter ε is kept fixed during the optimization process. After 4 iterates, the stopping criterion (5.4) is satisfied for $\eta = 10^{-4}$.

Figure 5.2a displays the convergence history of the gradient-based algorithm of Sect. 5.1. The evolution of the PDF of the random variable $\|y(z) - y_d\|_{L^2(D)}$ is plotted in Fig. 5.2b. The optimal risk-averse optimal control, solution to (5.3), is plotted in Fig. 5.3. The value of the objective function (5.2) at this risk-averse control is 0.3012.

Fig. 5.3 Optimal risk-averse
control, solution to (5.3)

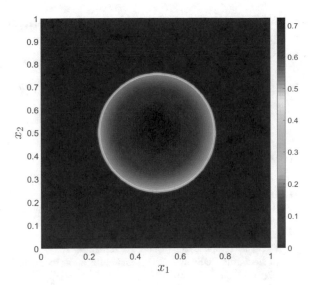

5.5   Notes and Related Software

The notion of risk aversion finds its origin in Economy [13]. This concept was then imported to Shape Optimization [5, 10] and to Control of Random PDEs (see [8, 9] and the references therein). Numerical resolution methods for this type of problems might be considered to be in its infancy. In this sense, the methods presented in this chapter are only a first step in this very challenging topic.

The risk functional considered in this chapter is only one among many other possible functionals measuring risk aversion (see e.g. [13, Chap. 2] and [5, 8, 11]).

Similarly to stochastic collocation methods, a primary drawback of stochastic Galerkin methods is the well known curse of dimensionality phenomenon, according to which the number of unknown coefficients highly increases as the space of random variables becomes higher. In this respect, optimality of ATD (and variants of it) in comparison with other possible choices for the index set Λ_g (ℓ, N), which determines the degrees of polynomials used for approximation in the random space, has been studied in [1, 2].

Breaking the curse of dimensionality in the computation of risk-averse functionals is a very challenging problem. The reader is referred to [4] for an interesting review on this topic.

For the different notions of convergence of sequences of random variables and their relationships, the reader is referred to [7].

A drawback of the algorithm presented in Sect. 5.3.3 is that the sample variance $\overline{\sigma}_{M_n}$ is used instead of the true variance σ, which is unknown. A higher order based algorithm has been proposed in [3].

In addition to the codes accompanying this book, open source codes for Stochastic Galerkin (or Polynomial Chaos) methods are provided in the following links:

- FERUM—http://www.ce.berkeley.edu/projects/ferum/ .
- OpenTURNS—http://www.openturns.org/.
- UQlab—http://www.uqlab.com/.
- Dakota software: https://dakota.sandia.gov/.
- Professor John Burkardt: https://people.sc.fsu.edu/jburkardt/.

References

1. Bäck, J., Nobile, F., Tamellini, L., Tempone, R.: Stochastic spectral Galerkin and colloca-tion methods for PDEs with random coefficients: a numerical comparison. Spectral and high order methods for partial differential equations, vol. 4362, Lect. Notes Comput. Sci. Eng., 76, Springer, Heidelberg (2011)
2. Bäck, J., Tempone, R., Nobile, F., Tamellini, L.: On the optimal polynomial approximation of stochastic PDEs by Galerkin and collocation methods. Math. Models Methods Appl. Sci. **22**(9), 1250023 (2012)
3. Bayer, C., Hoel, H., Schwerin, E.V., Tempone, R.: On non-asymptotic optimal stopping criteria in Monte Carlo simulations. SIAM J. Sci. Comput. **36**(2), 869–885 (2014)
4. Chen, P., Quarteroni, A.: Accurate and efficient evaluation of failure probability for partial differential equations with random input. Comput. Methods Appl. MEch. Engr. **267**, 233–260 (2013)
5. Conti, S., Held, H., Pach, M., Rumpf, M., Schultz, R.: Risk averse shape optimization. SIAM J. Control Optim. **49**(3), 927–947 (2011)
6. Fishman, G.S.: Monte Carlo: Concepts, Algorithms, and Applications. Springer, New York (1996)
7. Jacod, J., Protter, P.: Probability Essentials, 2nd edn. Springer, Berlin, Heidelberg (2003)
8. Kouri, D.P., Surowiec, T.M.: Risk-averse PDE-constrained optimization using the conditional value-at-risk. SIAM J. Optim. **26**(1), 365–396 (2016)
9. Marín, F.J., Martínez-Frutos, J., Periago, F.: A polynomial chaos approach to risk-averse piezo-electric control of random vibrations of beams. Int J Numer Methods Eng. 1–18 (2018). https://doi.org/10.1002/nme.5823
10. Martínez-Frutos, J., Herrero-Pérez, D., Kessler, M., Periago, F.: Risk-averse structural topology optimization under random fields using stochastic expansion methods. Comput. Methods Appl. Mech. Engr. **330**, 180–206 (2018)
11. Martínez-Frutos, J., Kessler, M., Periago, F.: Robust optimal shape design for an elliptic PDE with uncertainty in its input data. ESAIM: COCV, **21**(4), 901–923 (2015)
12. Nocedal, J., Wright, S.J.: Numerical Optimization. Springer Series in Operations Research. Springer, New York (1999)
13. Pflug, G.Ch., Römisch, W.: Modeling, Measuring, and Managing Risk. World Scientific Pub-lishing Co. Pte. Ltd. (2007)

Chapter 6
Structural Optimization Under Uncertainty

All human things hang on a slender thread, the strongest fall with a sudden crash.

Ovid.
(43 B.C-18 A.C)

Motivated by its applications in Engineering, since the early's 70, optimal design of structures has been a very active research topic. Roughly speaking, the structural optimization problem amounts to finding the optimal distribution of material within a design domain such that a certain objective function or mechanical criterion is minimized. To ensure robustness and reliability of the final designs, a number of uncertainties should be accounted for in realistic models. These uncertainties include, among others, manufacturing imperfections, unknown loading conditions and variations of material's properties.

This chapter aims at illustrating how the ideas and techniques elaborated in this text apply to the analysis and numerical resolution of the structural optimization problem under uncertainty. We would like to highlight that the goal of this chapter is not to provide a deep review of the techniques and results involved in structural optimization, but showing how the techniques described in the preceding chapters may be extended to the numerical resolution of the structural optimization problem under uncertainty. Most of the material of this chapter has been taken from [18, 19].

6.1 Problem Formulation

Let $(\Omega, \mathscr{F}, \mathbb{P})$ be a complete probability space, and let $\mathscr{O} \subset \mathbb{R}^d$ ($d = 2$ or 3 in applications) be a bounded Lipschitz domain whose boundary $\partial\mathscr{O}$ is decomposed into three disjoint parts

© The Author(s), under exclusive license to Springer Nature Switzerland AG 2018
J. Martínez-Frutos and F. Periago Esparza, *Optimal Control of PDEs under Uncertainty*,
SpringerBriefs in Mathematics, https://doi.org/10.1007/978-3-319-98210-6_6

$$\partial \mathcal{O} = \Gamma_D \cup \Gamma_N \cup \Gamma_0, \quad |\Gamma_D| > 0, \tag{6.1}$$

where $| \cdot |$ stands for the Lebesgue measure. Consider the system of linear elasticity with random input data

$$\begin{cases}
-\mathrm{div}(Ae(u(x, \omega))) = & f \quad \text{in } \mathcal{O} \times \Omega, \\
u(x, \omega) = & 0 \quad \text{on } \Gamma_D \times \Omega, \\
(Ae(u(x, \omega))) \cdot n = & g \quad \text{on } \Gamma_N \times \Omega, \\
(Ae(u(x, \omega))) \cdot n = & 0 \quad \text{on } \Gamma_0 \times \Omega,
\end{cases} \tag{6.2}$$

with $u = (u_1, \ldots, u_d)$ the displacement vector field, $e(u) = \left(\nabla u^T + \nabla u\right)/2$ the strain tensor and n the unit outward normal vector to $\partial \mathcal{O}$. The material Hooke's law A is defined for any symmetric matrix ζ by

$$A\zeta = 2\mu\zeta + \lambda(Tr\zeta)Id,$$

where $\lambda = \lambda(x, \omega)$ and $\mu = \mu(x, \omega)$ are the Lamé moduli of the material, which may depend on both $x = (x_1, \ldots, x_d) \in \mathcal{O}$ and $\omega \in \Omega$. Thus, $Ae(u)$ is the stress tensor. In the same vein, the volume and surface loads f and g depend on a spatial variable x and on a random event $\omega \in \Omega$, i.e. $f = f(x, \omega)$ and $g = g(x, \omega)$. The divergence (div) and the gradient ∇ operators in (6.2) involve only derivatives with respect to the spatial variable x.

Since the domain $\mathcal{O}$ will change during the optimization process, λ, μ and f must be known for all possible configurations of $\mathcal{O}$. Thus, a working bounded domain D is introduced, which contains all admissible domains $\mathcal{O}$ and satisfies that $\Gamma_D \cup \Gamma_N \subset \partial D$. Therefore, it is assumed that λ, μ and f are defined in D. Both Γ_D and Γ_N are kept fixed during optimization.

The following assumptions on the uncertain input parameters of system (6.2) are considered:

(A1) $\lambda, \mu \in L_{\mathbb{P}}^{\infty}(\Omega; L^{\infty}(D))$ and there exist $\mu_{min}, \mu_{max}, \lambda_{min}, \lambda_{max}$ such that

$$0 < \mu_{min} \leq \mu(x, \omega) \leq \mu_{max} < \infty \quad \text{a.e. } x \in D, \quad \text{a.s. } \omega \in \Omega,$$

and

$$0 < \lambda_{min} \leq 2\mu(x, \omega) + d\lambda(x, \omega) \leq \lambda_{max} < \infty \quad \text{a.e. } x \in D, \quad \text{a.s. } \omega \in \Omega,$$

(A2) $f = f(x, \omega) \in L_{\mathbb{P}}^2\left(\Omega; L^2(D)^d\right),$
(A3) $g = g(x, \omega) \in L_{\mathbb{P}}^2\left(\Omega; L^2(\Gamma_N)^d\right).$

Next, the Hilbert space

$$V_{\mathcal{O}} = \left\{v \in H^1(\mathcal{O})^d : v|_{\Gamma_D} = 0 \text{ in the sense of traces}\right\},$$

equipped with the usual $H^1(\mathcal{O})^d$-norm, is introduced.

Similarly to Sect. 3.1.1, a weak solution of (6.2) is a random field $u \in L^2_{\mathbb{P}}(\Omega; V_{\mathcal{O}})$ such that

$$\int_{\Omega} \int_{\mathcal{O}} Ae(u(x, \omega)) : e(v(x, \omega)) \, dx d\mathbb{P}(\omega) = \int_{\Omega} \left[\int_{\mathcal{O}} f \cdot v \, dx + \int_{\Gamma_N} g \cdot v \, ds \right] d\mathbb{P}(\omega),$$
$$(6.3)$$

for all $v \in L^2_{\mathbb{P}}(\Omega; V_{\mathcal{O}})$.

Notation. We recall that given two matrices $A = \left(a_{ij}\right)_{1 \le i, j \le d}$ and $B = \left(b_{ij}\right)_{1 \le i, j \le d}$, $A : B = \sum_{i,j=1}^{d} a_{ij} b_{ij}$ denotes de full contraction of A and B. If $u, v \in \mathbb{R}^d$, then $u \cdot v = \sum_{i=1}^{d} u_i v_i$ is the scalar product between vectors u and v.

By using Korn's inequality and following the same lines as in the deterministic case [1, Theorem 2.24], the well-posedness of (6.2) may be easily proved. Moreover, there exists $c > 0$ such that

$$\|u\|_{L^2_{\mathbb{P}}(\Omega; V_{\mathcal{O}})} \le c \left(\|f\|_{L^2_{\mathbb{P}}(\Omega; L^2(\mathcal{O})^d)} + \|g\|_{L^2_{\mathbb{P}}(\Omega; L^2(\Gamma_N)^d)} \right)$$

and

$$\|u(\omega)\|_{V_{\mathcal{O}}} \le c \left(\|f(\omega)\|_{L^2(\mathcal{O})^d} + \|g(\omega)\|_{L^2(\Gamma_N)^d} \right), \quad a.s. \quad \omega \in \Omega. \qquad (6.4)$$

Notice that, thanks to assumption (A1), the constant $c = c(\mathcal{O}, \mu_{min}, \lambda_{min})$ in (6.4) does not depend on $\omega \in \Omega$.

As is usual in structural optimization, a volume constraint is imposed on an admissible domain. Thus, the class of admissible domains is

$$\mathcal{U}_{ad} = \{\mathcal{O} \subset D, \quad \Gamma_D \cup \Gamma_N \subset \partial \mathcal{O}, \quad |\mathcal{O}| = V_0\}, \quad \text{with } 0 < V_0 < |D|.$$

For a given $\mathcal{O} \in \mathcal{U}_{ad}$, let $u_{\mathcal{O}} = u_{\mathcal{O}}(x, \omega)$ be the weak solution of (6.2). For each realization $\omega \in \Omega$, consider the compliance cost functional

$$J(\mathcal{O}, \omega) = \int_{\mathcal{O}} f(x, \omega) \cdot u_{\mathcal{O}}(x, \omega) \, dx + \int_{\Gamma_N} g(x, \omega) \cdot u_{\mathcal{O}}(y, \omega) \, ds. \qquad (6.5)$$

Similarly to Chap. 4, the following robust-type cost functional is introduced:

$$J_{R_\alpha}(\mathcal{O}) = \mathbb{E}(J(\mathcal{O}, \omega)) + \alpha \text{Var}(J(\mathcal{O}, \omega)), \qquad (6.6)$$

where $\alpha \ge 0$ is a weighting parameter,

$$\mathbb{E}(J(\mathcal{O}, \omega)) = \int_{\Omega} J(\mathcal{O}, \omega) \, d\mathbb{P}(\omega)$$

is the expectation of the compliance, and

$$\text{Var}(J(\mathcal{O},\omega)) = \int_{\Omega} J(\mathcal{O},\omega)^2 \, d\mathbb{P}(\omega) - \left(\int_{\Omega} J(\mathcal{O},\omega) \, d\mathbb{P}(\omega) \right)^2$$

is its variance.

Remark 6.1 Notice that in order for the variance to be well-defined we should assume that $f \in L^4_{\mathbb{P}}\left(\Omega; L^2(D)^d\right)$ and $g \in L^4_{\mathbb{P}}\left(\Omega; L^2(\Gamma_N)^d\right)$. Consequently, from now on in this chapter, whenever the variance is considered, it is assumed that $f \in L^4_{\mathbb{P}}\left(\Omega; L^2(D)^d\right)$ and $g \in L^4_{\mathbb{P}}\left(\Omega; L^2(\Gamma_N)^d\right)$.

Next, the following risk-averse (also called excess probability or failure probability in the context of structural optimization) objective function is considered:

$$J_{EP_\eta}(\mathcal{O}) = \mathbb{E}\left(H\left(J(\mathcal{O},\omega) - \eta\right)\right) \tag{6.7}$$

where $\eta > 0$ is a threshold parameter, and H is the Heaviside function.

With all these ingredients, the robust and risk-averse structural optimization problems are formulated as:

$$(P_\alpha) \quad \min\left\{J_{R_\alpha}(\mathcal{O}) \;:\; \mathcal{O} \in \mathscr{U}_{ad}\right\},$$

and

$$(P_{EP_\eta}) \quad \min\left\{J_{EP_\eta}(\mathcal{O}) \;:\; \mathcal{O} \in \mathscr{U}_{ad}\right\}.$$

Remark 6.2 Notice that a major difference of these two optimization problems, compared to the control problems studied in the preceding chapters, is that the optimization variable is not a function but a domain.

6.2 Existence of Optimal Shapes

If an admissible domain is assumed to be only Lebesgue measurable, then (P_α) and (P_{EP_η}) may be ill-posed due to the lack of closure of the set of admissible domains with respect to the weak-$\star$ topology in $L^\infty(D; \{0, 1\})$. To overcome this difficulty and so guarantee the existence of solution, additional conditions, such as smoothness, geometrical or topology constraints, may be imposed on the class of admissible designs. As an illustration of these ideas, the standard ε-cone property [5] is introduced next.

Let a point $y \in \mathbb{R}^d$, a vector $\xi \in \mathbb{R}^d$ and $\varepsilon > 0$ be given. Consider the cone

$$C(y, \xi, \varepsilon) = \left\{z \in \mathbb{R}^d : (z - y, \xi) \geq \cos\varepsilon |z - y|, \quad 0 < |z - y| < \varepsilon\right\}.$$

An open set $\mathcal{O} \subset \mathbb{R}^d$ is said to have the ε-cone property if for all $x \in \partial\mathcal{O}$ there exists a unit vector ξ_x such that $\forall y \in \overline{\mathcal{O}} \cap B(x, \varepsilon)$, the inclusion $C(y, \xi_x, \varepsilon) \subset \mathcal{O}$ holds.

Consider the class of domains

$$\mathcal{O}_\varepsilon = \{\mathcal{O} \text{ open}, \ \mathcal{O} \subset D, \quad \mathcal{O} \text{ has the } \varepsilon\text{-cone property}\}.$$

Eventually, from now on in this section, the class of admissible domains $\mathcal{U}_{ad}$ is replaced by

$$\mathcal{U}_{ad}^\varepsilon = \{\mathcal{O} \in \mathcal{O}_\varepsilon, \quad \Gamma_D \cup \Gamma_N \subset \partial\mathcal{O}, \quad |\mathcal{O}| = V_0\}, \quad \text{with } 0 < V_0 < |D|.$$

Following [13, Definition 2.2.3], let $\{\mathcal{O}_n\}_{n \geq 1}$ and $\mathcal{O}$ be (Lebesgue) measurable sets of $\mathbb{R}^d$. It is said that $\{\mathcal{O}_n\}$ converges to $\mathcal{O}$ in the sense of characteristic functions as $n \to \infty$ if

$$1_{\mathcal{O}_n} \longrightarrow 1_{\mathcal{O}} \quad \text{in } L_{loc}^p\left(\mathbb{R}^d\right) \quad \forall p \in [1, \infty[,$$

where 1_B stands for the characteristic function of a measurable set $B \subset \mathbb{R}^d$.

The class of domains $\mathcal{O}_\varepsilon$ enjoys the following compactness property: let $\{\mathcal{O}_n\}_{n \geq 1}$ be a sequence in the class $\mathcal{O}_\varepsilon$. Then, up to a subsequence, still labelled by n, $\mathcal{O}_n$ converges to some $\mathcal{O}^\star \in \mathcal{O}_\varepsilon$, in the sense of characteristic functions (also in the sense of Hausdorff distance and in the sense of compacts sets). We refer the readers to [13, Theorem 2.4.10] for more details on all these classes of convergence and their relationships.

Another very important property that the domains in the class $\mathcal{O}_\varepsilon$ satisfy is the following uniform extension property [5, Theorem II.1] : there exists a positive constant K, which depends only on ε, such that for all $\mathcal{O} \in \mathcal{O}_\varepsilon$ there exists a linear and continuous extension operator

$$P_{\mathcal{O}} : H^1\left(\mathcal{O}\right) \to H^1\left(\mathbb{R}^d\right), \quad \text{with } \|P_{\mathcal{O}}\| \leq K. \tag{6.8}$$

We are now in a position to prove the following existence result:

Theorem 6.1 *Problems* (P_α) *and* (P_{EP_n}) *have, at least, one solution.*

Proof Let $\mathcal{O}_n \subset \mathcal{U}_{ad}^\varepsilon$ be a minimizing sequence. Due to the compactness property of $\mathcal{O}_\varepsilon$ mentioned above there exists $\mathcal{O}^\star \in \mathcal{O}_\varepsilon$ and a subsequence, still labelled by n, such that $\mathcal{O}_n$ converges to $\mathcal{O}^\star$. Moreover, thanks to the convergence in the sense of characteristic functions, $\mathcal{O}^\star$ satisfies the volume constraint and hence $\mathcal{O}^\star \in \mathcal{U}_{ad}^\varepsilon$. The domain $\mathcal{O}^\star$ is the candidate to be the minimizer we are looking for.

For an arbitrary but fixed random event $\omega \in \Omega$ the following convergence holds:

$$J(\mathcal{O}_n, \omega) \to J(\mathcal{O}^\star, \omega) \text{ as } n \to \infty. \tag{6.9}$$

Indeed, let $u_n(x, \omega)$ and $u_{\mathcal{O}^\star}(x, \omega)$ be the solutions to (6.2) associated to $\mathcal{O}_n$ and $\mathcal{O}^\star$, respectively. Denoting by $\hat{u}_n(\cdot, \omega) := P_{\mathcal{O}_n}(u_n)(\cdot, \omega)$ the extension of $u_n(\cdot, \omega)$ to $\mathbb{R}^d$, by (6.4) and (6.8) one has

$$\|\hat{u}_n(\omega)\|_{V_D} \leq \|P_{\mathcal{O}_n}\| \|u_n(\omega)\|_{V_{\mathcal{O}_n}} \leq Kc\left(\|f(\omega)\|_{L^2(D)^d} + \|g(\omega)\|_{L^2(\Gamma_N)^d}\right),$$

a.s. $\omega \in \Omega$, which proves that $\hat{u}_n(\omega)$ is bounded in V_D. Thus, by extracting a sub-sequence, it converges weakly in V_D, and strongly in $L^2(D)^d$ to some $u^*(\omega) \in V_D$. Let us prove that

$$u^*|_{\mathcal{O}^*}(\omega) = u_{\mathcal{O}^*}(\omega), \quad \omega \in \Omega.$$

From the definition of u_n, it follows that

$$\int_{\mathcal{O}_n} Ae(u_n) : e(v)\, dx = \int_{\mathcal{O}_n} f \cdot v\, dx + \int_{\Gamma_N} g \cdot v\, dx \quad \forall v \in V_D, \text{ a.s. } \omega \in \Omega,$$

which is equivalent to

$$\int_D 1_{\mathcal{O}_n} Ae(\hat{u}_n) : e(v)\, dx = \int_D 1_{\mathcal{O}_n} f \cdot v\, dx + \int_{\Gamma_N} g \cdot v\, dx \quad \forall v \in V_D, \text{ a.s. } \omega \in \Omega.$$

The weak convergence of $\hat{u}_n$ in V_D and the strong convergence in $L^2(D)$ of the characteristic functions permit to take the limit leading to

$$\int_D 1_{\mathcal{O}^*} Ae(u^*) : e(v)\, dx = \int_D 1_{\mathcal{O}^*} f \cdot v\, dx + \int_{\Gamma_N} g \cdot v\, dx \quad \forall v \in V_D, \text{ a.s. } \omega \in \Omega,$$

and also, thanks to the extension property (6.8),

$$\int_{\mathcal{O}^*} Ae(u^*) : e(v)\, dx = \int_{\mathcal{O}^*} f \cdot v\, dx + \int_{\Gamma_N} g \cdot v\, dx, \quad \forall v \in V_{\mathcal{O}^*}.$$

This proves that $u^*|_{\mathcal{O}^*}(\omega) = u_{\mathcal{O}^*}(\omega)$ a.s. $\omega \in \Omega$. Since this is valid for any subsequence, the whole sequence $\hat{u}_n$ converges to $u_{\mathcal{O}^*}$. Let us consider the compliance in the form

$$J(\mathcal{O}_n, \omega) = \int_{\mathcal{O}_n} f \cdot u_n\, dx + \int_{\Gamma_N} g \cdot u_n\, ds = \int_D 1_{\mathcal{O}_n} f \cdot \hat{u}_n\, dx + \int_{\Gamma_N} g \cdot \hat{u}_n\, ds.$$

Since $1_{\mathcal{O}_n} f \to 1_{\mathcal{O}^*} f$ strongly in $L^2(D)$ and $\hat{u}_n \rightharpoonup u^*$ weakly in V_D, the limit in this expression can be taken to obtain (6.9).

Let us first consider the cost functional J_{R_α}. The two contributions of this cost functional are studied separately as follows:

- Expectation of the compliance: by the Cauchy-Schwartz inequality, the continuity of the trace operator and (6.4),

$$J(\mathscr{O}_n, \omega) = \int_{\mathscr{O}_n} f(x, \omega) \cdot u_{\mathscr{O}_n}(x, \omega)\, dx + \int_{\Gamma_N} g(x, \omega) \cdot u_{\mathscr{O}_n}(x, \omega)\, ds$$

$$\leq \|f(\omega)\|_{L^2(D)^d} \|u_{\mathscr{O}_n}(\omega)\|_{L^2(\mathscr{O}_n)^d} + \|g(\omega)\|_{L^2(\Gamma_N)^d} \|u_{\mathscr{O}_n}(\omega)\|_{L^2(\Gamma_N)^d}$$

$$\leq c\|f(\omega)\|_{L^2(D)^d}\left(\|f(\omega)\|_{L^2(D)^d} + \|g(\omega)\|_{L^2(\Gamma_N)^d}\right)$$

$$+ c\|g(\omega)\|_{L^2(\Gamma_N)^d}\left(\|f(\omega)\|_{L^2(D)^d} + \|g(\omega)\|_{L^2(\Gamma_N)^d}\right) \in L^1_{\mathbb{P}}(\Omega).$$

$$(6.10)$$

Combining (6.9) and (6.10), by dominated convergence, it follows that

$$\mathbb{E}(J(\mathscr{O}_n, \omega)) = \int_{\Omega} J(\mathscr{O}_n, \omega)\, d\mathbb{P}(\omega) \to \mathbb{E}(J(\mathscr{O}^\star, \omega)) = \int_{\Omega} J(\mathscr{O}^\star, \omega)\, d\mathbb{P}(\omega).$$

- Variance of the compliance: from (6.9) it follows that

$$J(\mathscr{O}_n, \omega)^2 \to J(\mathscr{O}^\star, \omega)^2 \text{ a.s. } \omega \in \Omega.$$

Hence, by Remark 6.1 and (6.10),

$$\int_{\Omega} J(\mathscr{O}_n, \omega)^2\, d\mathbb{P}(\omega) \to \int_{\Omega} J(\mathscr{O}^\star, \omega)^2\, d\mathbb{P}(\omega)$$

and

$$\text{Var}(J(\mathscr{O}_n, \omega)) = \int_{\Omega} J(\mathscr{O}_n, \omega)^2\, d\mathbb{P}(\omega) - \left(\int_{\Omega} J(\mathscr{O}_n, \omega)\, d\mathbb{P}(\omega)\right)^2 \to \text{Var}(J(\mathscr{O}^\star, \omega)).$$

Regarding the cost functional J_{EP_η}, since $F_n(\omega) := H(J(\mathscr{O}_n, \omega) - \eta)$ are integrable and non-negative, by Fatou's lemma,

$$\int_{\Omega} \liminf_{n\to\infty} F_n(\omega)\, d\mathbb{P}(\omega) \leq \liminf_{n\to\infty} \int_{\Omega} F_n(\omega)\, d\mathbb{P}(\omega). \qquad (6.11)$$

By (6.9) and the lower semi-continuity of the Heaviside function,

$$H\left(J(\mathscr{O}^\star, \omega) - \eta\right) \leq \liminf_{n\to\infty} F_n(\omega),$$

Integrating in both sides of this expression and using (6.11) yields

$$\int_{\Omega} H\left(J(\mathscr{O}^\star, \omega) - \eta\right) d\mathbb{P}(\omega) \leq \int_{\Omega} \liminf_{n\to\infty} F_n(\omega)\, d\mathbb{P}(\omega)$$

$$\leq \liminf_{n\to\infty} \int_{\Omega} F_n(\omega)\, d\mathbb{P}(\omega).$$

Since $\mathscr{O}_n$ is a minimizing sequence, the result follows.   □

6.3 Numerical Approximation via the Level-Set Method

In this section, a gradient-based method is used for the numerical approximation of problems (P_α) and (P_{EP_η}). Thus, the shape derivatives of the two considered cost functionals must be computed.

6.3.1 Computing Gradients of Shape Functionals

In his seminal paper [12], Hadamard introduced the concept of shape derivative of a functional with respect to a domain. This notion was further exploited in [20]. In this section, we briefly recall a few basic facts about shape derivatives, which are needed to compute the gradients of the shape functionals considered in this chapter. For further details on the notion of shape derivative and its properties we refer the reader to [1, 25].

We recall that the shape derivative of a functional $J : \mathcal{U}_{ad} \to \mathbb{R}$ at $\mathcal{O}$ is the Fréchet derivative at $\theta = 0$ in $W^{1,\infty}\left(\mathbb{R}^d; \mathbb{R}^d\right)$ of the mapping $\theta \mapsto J\left((I + \theta)(\mathcal{O})\right)$, i.e.,

$$J\left((I + \theta)(\mathcal{O})\right) = J(\mathcal{O}) + J'(\mathcal{O})(\theta) + o(\theta), \quad \text{with} \quad \lim_{\theta \to 0} \frac{|o(\theta)|}{\|\theta\|_{W^{1,\infty}}} = 0,$$

where $I : \mathbb{R}^d \to \mathbb{R}^d$ is the identity operator, $\theta \in W^{1,\infty}\left(\mathbb{R}^d; \mathbb{R}^d\right)$ is a vector field, and $J'(\mathcal{O})$ is a continuous linear form on $W^{1,\infty}\left(\mathbb{R}^d; \mathbb{R}^d\right)$.

The following result is very useful when it comes to calculating shape derivatives. For a proof, the reader is referred to [1, Proposition 6.22].

Proposition 6.1 *Let $\mathcal{O}$ be a smooth bounded open domain and let $\phi(x) \in W^{1,1}\left(\mathbb{R}^d\right)$. The shape functional*

$$J(\mathcal{O}) = \int_{\mathcal{O}} \phi(x)\, dx$$

is shape differentiable at $\mathcal{O}$ and its shape derivative is given by

$$J'(\mathcal{O})(\theta) = \int_{\partial\mathcal{O}} \theta(x) \cdot n(x)\, \phi(x)\, ds, \quad \theta \in W^{1,\infty}\left(\mathbb{R}^d; \mathbb{R}^d\right). \tag{6.12}$$

The volume constraint, as given in the definition of $\mathcal{U}_{ad}$, is incorporated in the cost functional through an augmented Lagrangian method. Accordingly, for the case of problem (P_α), the following augmented Lagrangian function is considered:

$$J^L_{R_\alpha}(\mathcal{O}) = J_{R_\alpha}(\mathcal{O}) + L_1\left(\int_{\mathcal{O}} dx - V_0\right) + \frac{L_2}{2}\left(\int_{\mathcal{O}} dx - V_0\right)^2. \tag{6.13}$$

For problem (P_{EP_η}), similarly to Chap. 5, the smooth approximation

$$J_{EP_\eta}^{L,\epsilon}(\mathcal{O}) = \int_\Omega \left(1 + e^{-\frac{2}{\epsilon}(J(\mathcal{O},\omega)-\eta)}\right)^{-1} d\mathbb{P}(\omega)$$
$$+ L_1 \left(\int_\mathcal{O} dx - V_0\right) + \frac{L_2}{2} \left(\int_\mathcal{O} dx - V_0\right)^2, \tag{6.14}$$

with $0 < \epsilon \ll 1$, is considered.

In both cases, $L = (L_1, L_2)$, being L_1 and L_2 the Lagrange multiplier and the penalty parameter, respectively, used to enforce the volume constraint at convergence. The Lagrange multiplier will be updated at each iteration n according to $L_1^{(n+1)} = L_1^n + L_2^n \left(\int_\mathcal{O} dx - V_0\right)$. The penalty multiplier L_2 will be augmented, after every five iterations, multiplying its previous value by 1.2.

Remark 6.3 Since during the optimization process Γ_D and Γ_N are kept fixed, the deformation vector field $\theta \in W^{1,\infty}\left(\mathbb{R}^d; \mathbb{R}^d\right)$ satisfies $\theta = 0$ on $\Gamma_D \cup \Gamma_N$.

Theorem 6.2 *The following assertions hold:*

(i) The shape derivative of the cost functional $J_{R_\alpha}^L(\mathcal{O})$ is given by

$$\left(J_{R_\alpha}^L\right)'(\mathcal{O})(\theta) = \int_{\Gamma_0} \theta \cdot n \left[\int_\Omega (2f \cdot u - Ae(u) : e(u)) \, d\mathbb{P}(\omega)\right] ds$$
$$+ \alpha \int_{\Gamma_0} \theta \cdot n \left[\int_\Omega (2J(\mathcal{O},\omega) f \cdot u - f \cdot p + Ae(u) : e(p)) \, d\mathbb{P}(\omega)\right] ds$$
$$- 2\alpha \mathbb{E}(J(\mathcal{O},\omega)) \int_{\Gamma_0} \theta \cdot n \left[\int_\Omega (2f \cdot u - Ae(u) : e(u)) \, d\mathbb{P}(\omega)\right] ds$$
$$+ L_1 \int_{\Gamma_0} \theta \cdot n \, ds + L_2 \left(\int_\mathcal{O} dx - V_0\right) \int_{\Gamma_0} \theta \cdot n \, ds \tag{6.15}$$

where $J(\mathcal{O},\omega)$ is given by (6.5) and the adjoint state $p \in L_\mathbb{P}^2(\Omega; V_\mathcal{O})$ solves the system

$$\int_\Omega \int_\mathcal{O} Ae(p) : e(v) \, dx d\mathbb{P}(\omega)$$
$$= -2 \int_\Omega \left(\int_\mathcal{O} f \cdot u \, dx + \int_{\Gamma_N} g \cdot u \, ds\right) \left(\int_\mathcal{O} f \cdot v \, dx + \int_{\Gamma_N} g \cdot v \, ds\right) d\mathbb{P}(\omega), \tag{6.16}$$

all $v \in L_\mathbb{P}^2(\Omega; V_\mathcal{O})$.

(ii) The shape derivative of $J_{EP_\eta}^{L,\epsilon}(\mathcal{O})(\theta)$ is expressed as

$$\left(J_{EP_\eta}^{L,\epsilon}\right)'(\mathcal{O})(\theta) = \int_{\Gamma_0} \theta \cdot n \left[\int_\Omega C_1(\omega) (2f \cdot u - Ae(u) : e(u)) \, d\mathbb{P}(\omega)\right] ds$$
$$+ L_1 \int_{\Gamma_0} \theta \cdot n \, ds + L_2 \left(\int_\mathcal{O} dx - V_0\right) \int_{\Gamma_0} \theta \cdot n \, ds, \tag{6.17}$$

where the random variable $C_1(\omega)$ is given by

$$C_1(\omega) = \frac{2}{\epsilon} \left(1 + e^{-\frac{2}{\epsilon}(J(\mathcal{O},\omega)-\eta)}\right)^{-2} e^{-\frac{2}{\epsilon}(J(\mathcal{O},\omega)-\eta)}. \tag{6.18}$$

Proof To simplify, a brief formal proof using the Lagrange method due to Céa [3] is provided next. We consider only the functional $J_{R_\alpha}^L$. The proof for $J_{EP_\eta}^{L,\epsilon}$ is similar on all points so that it is omitted here (see [19] for details).

The first term in the right hand side (r.h.s.) of (6.15), which is the one associated to the expectation of the compliance, is obtained, after changing the order of integration, from Proposition 6.1.

For the second term in (6.15), the Lagrangian

$$\mathscr{L}\left(\mathcal{O}, \hat{u}, \hat{p}\right) = \int_{\Omega} \left[\int_{\mathcal{O}} f \cdot \hat{u} \, dx + \int_{\Gamma_N} g \cdot \hat{u} \, ds \right]^2 d\mathbb{P}\left(\omega\right)$$
$$+ \int_{\mathcal{O}} \int_{\Omega} Ae\left(\hat{u}\right) : e\left(\hat{p}\right) d\mathbb{P}\left(\omega\right) dx \qquad (6.19)$$
$$- \int_{\mathcal{O}} \int_{\Omega} \hat{p} \cdot f \, d\mathbb{P}\left(\omega\right) dx$$
$$- \int_{\Gamma_N} \int_{\Omega} \hat{p} \cdot g \, d\mathbb{P}\left(\omega\right) ds$$

is considered, where $\hat{u}, \hat{p} \in L_{\mathbb{P}}^2(\Omega; H^1(\mathbb{R}^d; \mathbb{R}^d))$ satisfy $\hat{u} = \hat{p} = 0$ on Γ_D. For a given $\mathcal{O}$, denoting by (u, p) a stationary point of $\mathscr{L}$ and equating to zero the derivative of $\mathscr{L}$ with respect to $\hat{u}$ in the direction $v \in L_{\mathbb{P}}^2\left(\Omega; H^1(\mathbb{R}^d; \mathbb{R}^d)\right)$, with $v = 0$ on Γ_D, yields

$$0 = < \tfrac{\partial \mathscr{L}}{\partial \hat{u}}\left(\mathcal{O}, u, p\right), v >$$
$$= 2 \int_{\Omega} \left(\int_{\mathcal{O}} f \cdot u \, dx + \int_{\Gamma_N} g \cdot u \, ds \right) \left(\int_{\mathcal{O}} f \cdot v \, dx + \int_{\Gamma_N} g \cdot v \, ds \right) d\mathbb{P}(\omega)$$
$$+ \int_{\mathcal{O}} \int_{\Omega} Ae\left(v\right) : e\left(p\right) d\mathbb{P}\left(\omega\right) dx.$$

$$(6.20)$$

After integrating by parts and by taking v with compact support in $\mathcal{O} \times \Omega$ gives

$$- \operatorname{div}(Ae(p)) = -2 \left(\int_{\mathcal{O}} f \cdot u \, dx + \int_{\Gamma_N} g \cdot u \, ds \right) f \quad \text{in } D \times \Omega. \qquad (6.21)$$

Similarly, varying the trace of $v(\cdot, \omega)$ on Γ_N and on Γ_0, the following boundary conditions

$$Ae(p) \cdot n = -2 \left(\int_{\mathcal{O}} f \cdot u \, dx + \int_{\Gamma_N} g \cdot u \, ds \right) g \quad \text{on } \Gamma_N \times \Omega \qquad (6.22)$$

and

$$Ae(p) \cdot n = 0 \quad \text{on} \quad \Gamma_0 \times \Omega \qquad (6.23)$$

are obtained. The variational or weak formulation of (6.21)–(6.23) is given by (6.16). Notice that, from (6.16) it follows that

$$p = -2 \left(\int_{\mathcal{O}} f \cdot u \, dx + \int_{\Gamma_N} g \cdot u \, ds \right) u.$$

The third term in the r.h.s. of (6.15) is obtained by applying the chain rule theorem. The fourth term in (6.15) is obtained from Proposition 6.1. Recall that in the present case $\theta \cdot n = 0$ on $\Gamma_D \cup \Gamma_N$. Finally, the fifth term in (6.15) also follows from the chain rule theorem. □

6.3.2 *Mise en Scène of the Level Set Method*

Following the original idea by Osher and Sethian [22], which was imported to the topology optimization framework in [2, 27], the unknown domain $\mathcal{O}$ is described through the zero level set of a function $\psi = \psi(t, x)$ defined on the working domain D as follows

$$\forall x \in D, \ \forall t \in (0, T), \ \begin{cases} \psi(x, t) < 0 \text{ if } x \in \mathcal{O}, \\ \psi(x, t) = 0 \text{ if } x \in \partial \mathcal{O}, \\ \psi(x, t) > 0 \text{ if } x \notin \mathcal{O}. \end{cases} \quad (6.24)$$

Here, t stands for a fictitious time that accounts for the step parameter in the descent algorithm. Hence, the domain $\mathcal{O} = \mathcal{O}(t)$ is evolving in time during the optimization process. Let us assume that $\mathcal{O}(t)$ evolves with a normal velocity $V(t, x)$. Since

$$\psi(x(t), t) = 0 \quad \text{for any } x(t) \in \partial \mathcal{O}(t),$$

differentiating in t yields

$$\frac{\partial \psi}{\partial t} + \nabla \psi \cdot x'(t) = \frac{\partial \psi}{\partial t} + Vn \cdot \nabla \psi = 0.$$

Thus, as $n = \nabla \psi / |\nabla \psi|$,

$$\frac{\partial \psi}{\partial t} + V |\nabla \psi| = 0.$$

This *Hamilton-Jacobi* equation, which is complemented with an initial condition and with Neumann type boundary conditions, is solved in the working domain D since the velocity field is known everywhere in D. Indeed, returning to the optimization problems (P_α) and (P_{EP_η}), we recall that the shape derivatives of the cost functionals involved in those problems take the form

$$\int_{\Gamma_0} v\theta \cdot n \, ds$$

where

$$\begin{aligned} v = &\int_\Omega (2f \cdot u - Ae(u) : e(u)) \, d\mathbb{P}(\omega) \\ &+ \alpha \int_\Omega (2J(\mathcal{O}, \omega) f \cdot u - f \cdot p + Ae(u) : e(p)) \, d\mathbb{P}(\omega) \\ &- 2\alpha \mathbb{E}(J(\mathcal{O}, \omega)) \int_\Omega (2f \cdot u - Ae(u) : e(u)) \, d\mathbb{P}(\omega) \\ &+ L_1 + L_2 \left(\int_{\mathcal{O}} dx - V_0 \right) \end{aligned} \quad (6.25)$$

for the case of $J_{R_\alpha}^L$, and

$$v = \int_{\Omega} C_1(\omega) (2f \cdot u - Ae(u) : e(u)) \, d\mathbb{P}(\omega)$$

$$+ L_1 + L_2 \left(\int_{\mathscr{O}} dx - V_0 \right), \tag{6.26}$$

for $J_{EP_\eta}^{L,\epsilon}$ (see Theorem 6.2). Since u and p are computed in D, the integrand in v is defined also in D and not only on the free boundary Γ_0. As a consequence, a descent direction in the whole domain D is defined by

$$\theta = -vn.$$

Therefore, its normal component $\theta \cdot n = -v$ is the advection velocity in the Hamilton-Jacobi equation

$$\frac{\partial \psi}{\partial t} - v|\nabla \psi| = 0, \ t \in (0, T), \quad x \in D. \tag{6.27}$$

As indicated above, Eq. (6.27) is solved in D. To this end, the elasticity system (6.2) as well as the adjoint system (6.16) are solved in D by using the so-called *ersatz material* approach. Precisely, $D \setminus \mathscr{O}$ is filled by a weak phase that mimics the void, but at the same time avoids the singularity of the stiffness matrix. Thus, an elasticity tensor $A^*(x)$ which equals A in $\mathscr{O}$ and $10^{-3} \cdot A$ in $D \setminus \mathscr{O}$ is introduced. The displacement field $u(x, \omega)$ then solves

$$\begin{cases} -\text{div}(A^* e(u(x, \omega))) = f & \text{in } D \times \Omega, \\ u(x, \omega) = 0 & \text{on } \Gamma_D \times \Omega, \\ (A^* e(u(x, \omega))) \cdot n = g & \text{on } \Gamma_N \times \Omega, \\ (A^* e(u(x, \omega))) \cdot n = 0 & \text{on } \partial D_0 \times \Omega, \end{cases} \tag{6.28}$$

Equation (6.27) may be solved by using an explicit first order upwind scheme. To avoid singularities, the level-set function is periodically reinitialized. More details on these implementation issues of the level set are provided in Sect. 6.5.

A crucial step is the numerical approximation of the advection velocities (6.25) and (6.26). For the case of problem (P_α), this may be done by using the Stochastic Collocation method presented in Chap. 4. For problem (P_{EP_η}), the techniques of Chap. 5 may be applied. See [18, 19] for more details.

6.4 Numerical Simulation Results

This section presents numerical simulation results for problems (P_α) and (P_{EP_η}) in the widely studied benchmark example of the beam-to-cantilever.

The beam-to-cantilever problem consists in the shape optimization of a two-dimensional cantilever under uncertain loading conditions. The left edge of the cantilever is anchored and a unit force with uncertainty in direction, centered in

Fig. 6.1 The beam-to-cantilever problem: (a) the design domain and boundary conditions, and (b) the initialization of level-set function. From [18], reproduced with permission

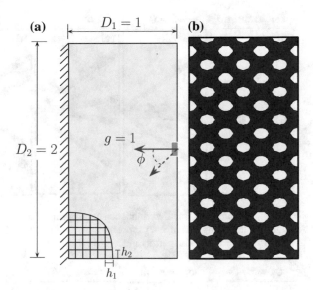

the horizontal line, is applied at the middle of the right edge. The design domain is a 1×2 rectangle which is tessellated using a 60×120 quadrangular mesh ($h_1 = h_2 = 0.016$). The uncertain loading, the boundary conditions and the tessellation of the design domain are depicted in Fig. 6.1a, whereas the initialization of the level-set function is shown in Fig. 6.1b.

All the experiments that follow make use of a fixed grid of $\mathbb{Q}_1$ quadrangular plane stress finite elements with Young's modulus $E = 1$ and Poisson's ratio $\nu = 0.3$. The level set function $\psi(t, x)$ is discretized at the nodes of such a fixed grid. For J_{R_α}, the Lagrange multipliers and the penalty parameters of the augmented Lagrangian functions are initialized heuristically according to $L_1 = 0.1 \cdot J_{R_\alpha}(\mathcal{O})/V_0$ and $L_2 = 0.01 \cdot L_1/V_0$. The target volume is set to 30% of the initial design domain. The same strategy is followed for $J_{EP_\eta}^{L,\epsilon}$.

Problem (P_α). The direction ϕ of the unit-load follows a Gaussian distribution centered at the horizontal line, $\phi = 0$, with the deterministic loading state as the mean value $\mu_\phi = 0$. The influence of the level of uncertainty on the robust shape design is shown by the consideration of $\sigma_\phi = \{\pi/6, \pi/12, \pi/24, \pi/32, \pi/42\}$ standard deviation values.

The deterministic shape optimization design (classical beam) for the horizontal ($\phi = 0$) unit-load is depicted in Fig. 6.2a. The robust optimized designs for the minimization of the cost functional $J_{R_\alpha}^L$ are shown from Fig. 6.2b to Fig. 6.2e for various magnitudes of the parameter α. The resulting robust designs incorporate stiffness in the vertical direction where the uncertain loading is introduced. Higher values of the parameter α, increasing the weight of the variance in the functional, lead to more robust structural designs at the cost of decreasing their (compliance) optimality. One can observe how the increase in robustness is firstly achieved by

Deterministic $\alpha = 0$ $\alpha = 1$ $\alpha = 2$ $\alpha = 3$

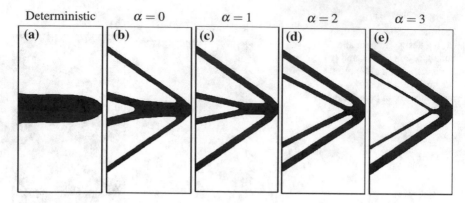

Fig. 6.2 Problem (P_α). **(a)** Deterministic shape optimization design and **(b–e)** robust optimized designs for the beam-to-cantilever problem with the direction of the unit-load $\phi \sim \mathcal{N}(0, \pi/12)$. From [18], reproduced with permission

$\sigma_\phi = \pi/42$ $\sigma_\phi = \pi/32$ $\sigma_\phi = \pi/24$ $\sigma_\phi = \pi/12$ $\sigma_\phi = \pi/6$

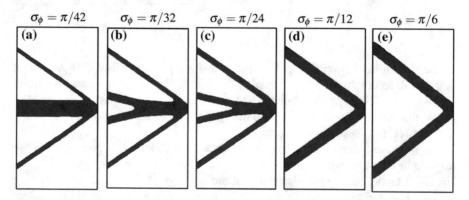

Fig. 6.3 Problem (P_α). Robust optimized designs for $J_{R_\alpha}^L$, with $\alpha = 4$ increasing the uncertainty in the direction of the unit-load $\phi \sim \mathcal{N}(0, \sigma_\phi)$. From [18], reproduced with permission

means of topology modifications, increasing the redundancy by adding paths to transmit the load to the support, and then by thickening the cantilever members.

The effect of the amount of uncertainty using a Gaussian probability distribution for the direction ϕ of the unit-load is shown in Fig. 6.3. This experiment shows the different shape optimal designs of the functional cost $J_{R_\alpha}^L$ for a given value of α. One can observe how the increment in the uncertainty of the direction ϕ of the unit-load increases the span between supports and the thickness of members, which increment the vertical stiffness.

Problem (P_{EP_η}). The optimal designs are shown from Fig. 6.4a to Fig. 6.4e for various magnitudes of the parameter η. One can observe that the resulting designs incorporate stiffness in the vertical direction where the uncertain loading is introduced. Analogously to what was observed in problem (P_α), the risk is firstly minimized

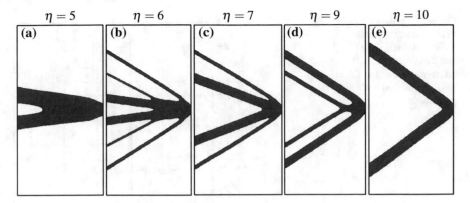

Fig. 6.4 Problem (P_{EP_η}). The beam-to-cantilever problem: Optimal designs for different η values. From [19], reproduced with permission

Table 6.1 Problem (P_{EP_η}). Excess probability of optimal design for different values of η and for the deterministic optimal design. From [19], reproduced with permission

Case	$\mathbb{P}\{J > 5\}$	$\mathbb{P}\{J > 6\}$	$\mathbb{P}\{J > 7\}$	$\mathbb{P}\{J > 9\}$	$\mathbb{P}\{J > 10\}$	σ_J
Deterministic	0.7524	0.6269	0.5419	0.4224	0.3763	9.2215
$\eta = 5$	**0.7093**	0.5274	0.4155	0.2772	0.2280	5.4176
$\eta = 6$	0.9999	**0.4632**	0.2884	0.1299	0.0872	2.3927
$\eta = 7$	1	0.4861	**0.2178**	0.0614	0.0311	1.3829
$\eta = 9$	1	1	0.6557	**0.0581**	0.0249	0.8515
$\eta = 10$	1	1	1	0.1093	**0.0225**	0.5725

by means of topology modifications, increasing the redundancy by adding paths to transmit the load to the support, and then by thickening the cantilever members.

Table 6.1 shows the probability of exceeding different magnitudes of compliance for deterministic and risk-averse designs. Notice that the deterministic design is not always the worst choice when compared to risk averse designs with $\eta \neq \eta_i$. This is attributed to the shape of the PDF resulting from the minimization of the excess probability. Figure 6.5 shows the PDF for the optimal deterministic design and diverse optimal risk-averse designs using different values of η. One can observe that the mean values of these PDFs are higher as the prescribed threshold η_i increases. Although the probability of exceeding η_i is reduced, more realizations of the compliance are close to such a threshold value, and thus the mean performance is degraded. This can lead to a deterministic design showing lower levels of $\mathbb{P}\{J > \eta_i\}$ compared to risk-averse designs obtained with $\eta > \eta_i$ values. This conclusion is only pertinent to the problem discussed in this specific example. Finally, it should be noted that the reduction in the excess probability involves a reduction of the dispersion in the PDF. This fact can be observed in the last column of Table 6.1, where the standard deviation of the compliance is shown for the different cases analyzed. One can observe that the value of the standard deviation is reduced as the threshold η increases. In addition, to

Fig. 6.5 Problem (P_{EP_η}).
PDFs calculated for the
optimal deterministic design
and risk-averse designs with
different values of η. From
[19], reproduced with
permission

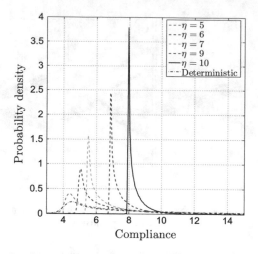

reduce the excess probability, the risk-averse formulation provides designs which are less sensitive to fluctuations and thus more robust. This fact explains the topological similarities between the risk-averse designs (Fig. 6.4) and the robust designs obtained in problem (P_α) above (Fig. 6.2).

6.5   Notes and Related Software

Structural optimization provides engineering designers with a powerful tool to find innovative and high-performance conceptual designs at the early stages of the design process. Such a technique has been successfully applied to improve the design of complex industrial problems, such as aeronautical, aerospace and naval applications [10]. The need for including uncertainty during the design process has shown to be a key issue for solving real-world engineering problems in several fields, such as aeronautical and aerospace [17], civil [14], automotive [26] and mechanical [23] engineering, to name but a few.

More recently, the structural optimization problem including uncertainties in loads, material properties and even in geometry has been studied by using both the SIMP method (see [16] and the references therein) and the level set method (see, e.g. [4, 7–9, 11]). At the discrete level, it is also remarkable to mention [6].

A previous knowledge of Hadamard's boundary variation method and of the level set method has been assumed in this chapter. For a comprehensive treatment of these methods we refer the reader to [1, 21, 24]. Section 6.3.2 lacks many numerical details and implementation tricks, which are out of the scope of this text. The interested reader may find open source codes for the numerical resolution of the *deterministic* structural optimization problem by using the level set method in the two following links:

- Several very useful codes written in Scilab and FreeFem++ are provided by Professor G. Allaire. http://www.cmap.polytechnique.fr/ allaire/
- A level set-based structural optimization educational code using FEniCS is provided by A. Laurain in [15]. http://www.antoinelaurain.com/

References

1. Allaire, G.: Conception Optimale de Structures, 58. Mathématiques et Applications. Springer, Berlin (2007)
2. Allaire, G., Jouve, F., Toader, A.M.: Structural optimization using shape sensitivity analysis and a level-set method. J. Comput. Phys. **194**, 363–393 (2004)
3. Céa, J.: Conception optimale ou identification de formes, calcul rapide de la dérivée directionnelle de la function cöut. ESAIM Math. Model. Numer. Anal. **20**, 371–402 (1986)
4. Chen, S., Chen, W., Lee, S.: Level set based robust shape and topology optimization under random field uncertainties. Struct. Multidisc. Optim. **44**(1), 507–524 (2010)
5. Chenais, D.: On the existence of a solution in a domain identification problem. J. Math. Anal. Appl. **52**, 189–219 (1975)
6. Choi, S.K., Grandhi, R., Canfield, R.A.: Reliability-Based Structural Design, Springer (2007)
7. Conti, S., Held, H., Pach, M., Rumpf, M., Schultz, R.: Risk averse shape optimization. SIAM J. Control Optim. **49**, 927–947 (2011)
8. Dambrine, M., Dapogny, C., Harbrecht, H.: Shape optimization for quadratic functionals and states with random right-hand sides. SIAM J. Control Optim. **53**(5), 3081–3103 (2015)
9. Dambrine, M., Harbrecht, H., Puig, B.: Computing quantities of interest for random domains with second order shape sensitivity analysis. ESAIM Math. Model. Numer. Anal. **49**(5), 1285–1302 (2015)
10. Deaton, J.D., Grandhi, R.V.: A survey of structural and multidisciplinary continuum topology optimization: post 2000. Struct. Multidiscip. Optim. **49**, 1–38 (2014)
11. Dunning, P.D., Kim, H.A.: Robust topology optimization: minimization of expected and variance of compliance. AIAA J. **51**(11), 2656–2664 (2013)
12. Hadamard, J.: Mémoire sur le problème d'analyse relatif l'équilibre des plaques élastiques encastrées. Bull. Soc. Math, France (1907)
13. Henrot, A., Pierre, M.: Variation et optimisation de formes: Une Analyse Geometrique, 48. Mathématiques et Applications. Springer, Berlin (2005)
14. Lagaros, N.D., Papadrakakis, M.: Robust seismic design optimization of steel structures. Struct. Multidiscip. Optim. **33**(6), 457–469 (2007)
15. Laurain, A.: A level set-based structural optimization code using FEniCS (2018). arXiv:1705.01442
16. Lazarov, B., Schevenels, M., Sigmund, O.: Topology optimization considering material and geometric uncertainties using stochastic collocation methods. Struct. Multidisc. Optim. **46**(4), 597–612 (2012)
17. López, C., Baldomir, A., Hernández, S.: The relevance of reliability-based topology optimization in early design stages of aircraft structures. Struct. Multidiscip. Optim. **57**(1), 417–439 (2018)
18. Martínez-Frutos, J., Herrero-Pérez, D., Kessler, M., Periago, F.: Robust shape optimization of continuous structures via the level set method. Comput. Methods Appl. Mech. Engrg. **305**, 271–291 (2016)
19. Martínez-Frutos, J., Herrero-Pérez, D., Kessler, M., Periago, F.: Risk-averse structural topology optimization under random fields using stochastic expansion methods. Comput. Methods Appl. Mech. Engrg. **330**, 180–206 (2018)
20. Murat, F., Simon, J.: Sur le contrôle par un domaine géométrique, Technical report RR-76015, Laboratoire d'Analyse Numérique (1976)

21. Osher, S., Fedkiw, R.: Level Set Methods and Dynamic Implicit Surfaces. Springer, New York (2003)
22. Osher, S., Sethian, J.A.: Fronts propagation with curvature-dependent speed: algorithms based on Hamilton-Jacobi formulations. J. Comput. Phys. **79**(1), 12–49 (1988)
23. Schevenelsa, M., Lazarov, B.S., Sigmund, O.: Robust topology optimization accounting for spatially varying manufacturing errors. Comput. Methods Appl. Mech. Engrg. **200**, 3613–3627 (2011)
24. Sethian, J.A.: Level Set Methods and Fast Marching Methods: Evolving Interfaces in Computational Geometry, Fluid Mechanics, Computer Vision and Material Science. Cambridge University Press (1999)
25. Sokolowski, J., Zolesio, J. P.: Introduction to Shape Optimization: Shape Sensitivity Analysis. Springer Series in Computational Mathematics, vol. 16, Springer, Heidelberg (1992)
26. Youn, B.D., Choi, K.K., Yang, R.J., Gu, L.: Reliability-based design optimization for crashworthiness of vehicle side impact. Struct. Multidiscip. Optim. **26**(3–4), 272–283 (2004)
27. Wang, M.Y., Wang, X., Guo, D.: A level-set method for structural topology optimization. Comput. Methods Appl. Mech. Engrg. **192**, 227–246 (2003)

Chapter 7
Miscellaneous Topics and Open Problems

*Surely the first and oldest problems in every branch of
mathematics spring from experience and are suggested by the
world of external phenomena.*

David Hilbert.
Lecture delivered before ICM, Paris, 1900.

In this final chapter we discuss some related topics to the basic methodology presented
in detail in the previous chapters. These topics are related to (i) time-dependent
problems, and (ii) physical interpretation of robust and risk-averse optimal controls.
We also list a number of challenging problems in the field of control under uncertainty.

7.1 The Heat Equation Revisited II

In this section it is shown how to apply the techniques developed in the preceding
chapters to robust optimal controls problems constrained by a time-dependent PDE.
As it will be shown hereafter, the methods described in Chap. 4 easily extend to this
situation.

As an illustration, uncertainty will not appear in the principal part of the differential
operator (as it was the case in the examples of the previous chapters), but in an input
parameter acting on the boundary conditions. The control function will act in a part
of the boundary's domain. To this end, the heat equation

$$
\begin{cases}
y_t - \Delta y = 0, & \text{in } (0, T) \times D \times \Omega \\
\nabla y \cdot n = 0, & \text{on } (0, T) \times \partial D_0 \times \Omega \\
\nabla y \cdot n = \alpha \, (u - y), & \text{on } (0, T) \times \partial D_1 \times \Omega, \\
y\,(0) = y^0 & \text{in } D \times \Omega,
\end{cases}
\tag{7.1}
$$

is revisited again.

© The Author(s), under exclusive license to Springer Nature Switzerland AG 2018
J. Martínez-Frutos and F. Periago Esparza, *Optimal Control of PDEs under Uncertainty*,
SpringerBriefs in Mathematics, https://doi.org/10.1007/978-3-319-98210-6_7

It is well-known [2, 5] that the convective heat transfer coefficient α is difficult to measure in practice as it depends on random factors such as the physical properties of the boundary ∂D_1 and the motion of the surrounding fluid. Thus, it is reasonable to assume that $\alpha = \alpha(x, \omega)$ is a random field.

For a given control time $T > 0$ and a target function $y_d \in L^2(D)$, the two following cost functionals are introduced:

$$J_1(u) = \tfrac{1}{2} \int_D \int_\Omega |y(T, x, \omega) - y_d(x)|^2 \, d\mathbb{P}(\omega) dx$$

$$+ \tfrac{\gamma_1}{2} \int_D Var\,(y(T, x)) \, dx \tag{7.2}$$

$$+ \tfrac{\gamma_2}{2} \int_0^T \int_{\partial D_1} \left(|u(t, y)|^2 + |u_t(t, y)|^2 \right) ds \, dt$$

and

$$J_2(u) = \int_D \left(\int_\Omega y(T, x, \omega) \, d\mathbb{P}(\omega) - y_d(x) \right)^2 dx$$

$$+ \tfrac{\gamma_1}{2} \int_D Var\,(y(T, x)) \, dx \tag{7.3}$$

$$+ \tfrac{\gamma_2}{2} \int_0^T \int_{\partial D_1} \left(|u(t, y)|^2 + |u_t(t, y)|^2 \right) ds \, dt.$$

In both cases, $\gamma_1, \gamma_2 \geq 0$ are two weighting parameters, and the space of admissible controls is $\mathscr{U}_{ad} = H_0^1\left(0, T; L^2(\partial D_1)\right)$.

With all these ingredients, the two following optimal control problems are considered:

$$(P_{heat}) \begin{cases} \text{Minimize in } u \in \mathscr{U}_{ad} : J_1(u) \quad \text{or} \quad J_2(u) \\ \text{subject to} \\ \qquad\qquad\qquad y = y(u) \text{ solves (7.1).} \end{cases}$$

To fix ideas, in what follows, $D = (0, 1)^2$ is the unit square. Its boundary is divided into $\partial D_1 = \{(x_1, x_2) \in \mathbb{R}^2 : x_2 = 0\} \cup \{(x_1, x_2) \in \mathbb{R}^2 : x_2 = 1\}$ and $\partial D_0 = \partial D \setminus \partial D_1$.

Since the random field α is positive, it is typically modeled as a log-normal random field. For the experiment that follows, α is a log-normal random field with mean equal to one and variance 0.01 at $x_2 = 0$, and variance 0.1 at $x_2 = 1$. Precisely, in a first step, a three-terms, $z_3(x, \omega)$, KL expansion is computed using the exponential covariance function (2.15), with correlation length of 0.5. In a second step, formulae (2.27), with $\bar{a} = 1$ and variances 0.01 at $x_2 = 0$, and variance 0.1 at $x_2 = 1$, are used to compute the parameters μ and σ that appear in (2.26) with $z(x, \omega)$ replaced by $z_3(x, \omega)$. This leads to an explicit representation of $\alpha(x, \omega)$ as given by (2.26) with $a(x, \omega)$ replaced by $\alpha(x, \omega)$. Note that two three-terms KL expansions are considered; one of them for the edge $x_2 = 0$ and another one for $x_2 = 1$. Thus, we end up with 6 stochastic directions.

Fig. 7.1 Problem (P_{heat}).
Deterministic target function
y_d. From [11], reproduced
with permission

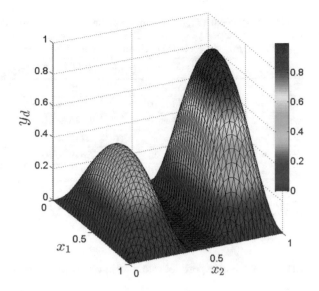

Similarly to Chap. 4, the random domain $\Gamma = [-3, 3]^6$, and

$$\rho(z) = \prod_{n=1}^{6} \phi(z_n), \quad z = (z_1, \cdots, z_6) \in \Gamma, \tag{7.4}$$

where $\phi(z_n)$ is the truncated Gaussian (2.30).

The initial condition is $y(0, x, \omega) = 0$, the final control time is $T = 0.5$, and the target function is (see Fig. 7.1)

$$y_d(x_1, x_2) = \begin{cases} (-2x_1^2 + 2x_1)(-25x_2^2 + 10x_2) & 0 \le x_1 \le 1, \ 0 \le x_2 \le 0.4, \\ 0 & 0 \le x_1 \le 1, \ 0.4 \le x_2 \le 0.6, \\ (-4x_1^2 + 4x_1)(-25x_2^2 + 40x_2 - 15) & 0 \le x_1 \le 1.0, \ 0.6 \le x_2 \le 1. \end{cases} \tag{7.5}$$

The gradient-based minimization algorithm of Sect. 4.2 may be used to solve problem (P_{heat}). Explicit expressions for the descent directions and the optimal step-size parameters are provided in [11]. Here, the only novelty is that the discretization process, which is needed to approximate numerically the state equation (7.1) and its associated adjoint equations, involves discretization w.r.t. the time variable. This may be done, for instance, by using a fully implicit backward second order Gear scheme. Indeed, starting from $y^{(n-1)}$ at time $t^{(n-1)}$ and $y^{(n)}$ at time $t^{(n)}$, the Gear scheme is based on the approximation

$$y_t^{(n+1)} \equiv \frac{\partial y}{\partial t}(t^{(n+1)}) \simeq \frac{3y^{(n+1)} - 4y^{(n)} + y^{(n-1)}}{2dt}. \tag{7.6}$$

Table 7.1 Problem (P_{heat}). Summary of the error with respect to the target and variance of the state variable for $\gamma_2 = 1e - 8$. Here $\bar{y}(T, x)$ denotes the mean of $y(T, x, \cdot)$, i.e., $\bar{y}(T, x) = \int_\Gamma y(T, x, z)\rho(z)\, dz$. From [11], reproduced with permission

	$\|\bar{y}(T) - y_d\|^2_{L^2(D)}$	$\|y(T) - y_d\|^2_{L^2(D)\otimes L^2_\rho(\Gamma)}$	$\|Var(y(T))\|_{L^2(D)}$
$J_1, \gamma_1 = 0$	8.669×10^{-3}	1.086×10^{-2}	2.187×10^{-3}
$J_1, \gamma_1 = 1$	6.330×10^{-3}	8.076×10^{-3}	1.745×10^{-3}
$J_1, \gamma_1 = 2$	5.945×10^{-3}	7.466×10^{-3}	1.519×10^{-3}
$J_1, \gamma_1 = 3$	7.388×10^{-3}	1.138×10^{-3}	8.526×10^{-3}
$J_2, \gamma_1 = 0$	1.128×10^{-3}	7.649×10^{-3}	6.521×10^{-3}
$J_2, \gamma_1 = 1$	1.969×10^{-3}	3.283×10^{-3}	1.314×10^{-3}
$J_2, \gamma_1 = 2$	1.867×10^{-3}	8.559×10^{-4}	2.723×10^{-3}
$J_2, \gamma = 3$	1.738×10^{-3}	7.294×10^{-4}	2.468×10^{-3}

In the numerical experiments that follow, a time step of $dt = 2.5 \times 10^{-3}$ is considered in (7.6).

Discretization in space is carried out by using Lagrange $P1$ finite elements on a triangular mesh composed of 2436 triangles and 1310 degrees of freedom.

Following Sect. 4.2.2.1, an anisotropic sparse grid is constructed for discretization in the random domain Γ. The same vector of weights as in (4.25) has been selected. See [12] for more details on this passage. The level of quadrature ℓ in the sparse grid is fixed by using the adaptive algorithm of Sect. 4.2.2.2. Precisely, a non-nested quadrature rule whose collocation nodes are determined by the roots of Hermite polynomials is used. Notice that the random variables of this problem follow the Gaussian distribution (7.4) so that Hermite polynomials are a suitable choice for approximation in the random domain. The nodes and weights for the anisotropic sparse grid are computed adaptively as described in Sect. 4.2.2 with $N = 6$, $\bar{\ell} = 9$, $\delta = 1e - 6$ and $\ell_{opt} = 5$. Here, the only difference is that in the stopping criteria (4.38) and (4.40), $y = y(T)$ is the solution to (7.1) at time T.

The convergence history of the optimization algorithm, depicted in Fig. 7.2, confirms that the cost J_2 requires more iterations to achieve convergence.

To analyze the performance of the cost functionals J_1 and J_2, the optimization problem has been solved for values of $\gamma_1 = 0$, 1, 2, 3, and $\gamma_2 = 1e - 8$. The results for the different contributions of the costs J_1 and J_2 are presented numerically in Table 7.1 and graphically in Figs. 7.3, 7.4 and 7.5. As expected, the peak value of the variance of the state variable is close to the upper edge due to the large covariance of the random field on this part of the boundary. The results presented in Figs. 7.3 and 7.4 are in line with the ones obtained in Chap. 4: on the one hand, the cost J_1 provides solutions less sensitive to variations in the input data. On the other hand, the cost J_2 provides a better correspondence between the mean of state variable and the target function, but with high variance. This variance can be reduced by increasing the value of the parameter γ_1 as shown in Fig. 7.5 for $\gamma_1 = 1$, 2 and 3.

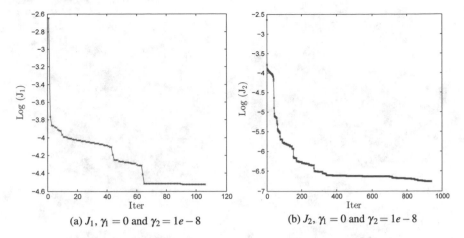

(a) J_1, $\gamma_1 = 0$ and $\gamma_2 = 1e - 8$ (b) J_2, $\gamma_1 = 0$ and $\gamma_2 = 1e - 8$

Fig. 7.2 Problem (P_{heat}). Convergence history of the gradient-based minimization algorithm. From [11], reproduced with permission

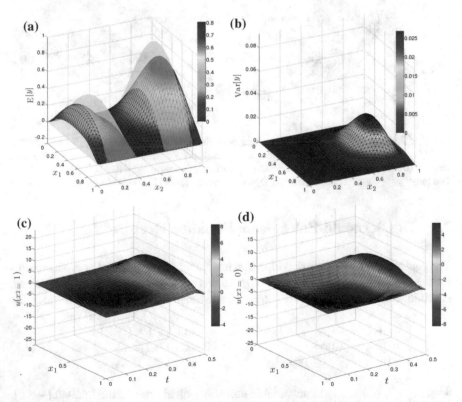

Fig. 7.3 Problem (P_{heat}). Results associated with the cost J_1 with $\gamma_1 = 0$, $\gamma_2 = 1e - 8$. The following quantities are represented: the mean of the state variable (**a**), the variance of the state variable (**b**), the control variable acting on $x_2 = 1$ (**c**), and the control variable acting on $x_2 = 0$ (**d**). The target y_d is shown transparently in (**a**) as a reference. From [11], reproduced with permission

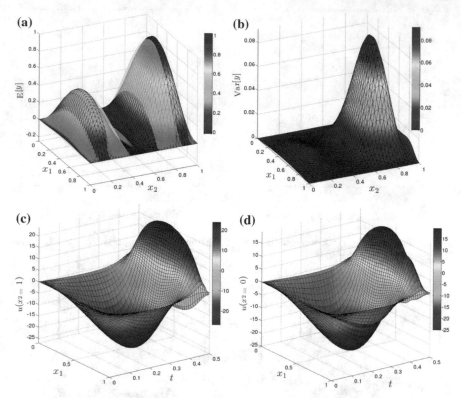

Fig. 7.4 Problem (P_{heat}). Results associated with the cost J_2 with $\gamma_1 = 0$, $\gamma_2 = 1e - 8$. The following quantities are represented: the mean of the state variable (**a**), the variance of the state variable (**b**), the control variable acting on $x_2 = 1$ (**c**), and the control variable acting on $x_2 = 0$ (**d**). The target y_d is shown transparently in (**a**) as a reference. From [11], reproduced with permission

7.2 The Bernoulli-Euler Beam Equation Revisited II

The goal of this section is to provide a first insight into the profiles of optimal controls for the class of robust optimal control problems considered in this text. To this end, the following control system for the Bernoulli-Euler beam equation is considered:

$$\begin{cases} y_{tt} + (1 + \alpha)\, y_{xxxx} = 1_{\mathscr{O}} u(t, x), & \text{in } (0, T) \times (0, L) \times \Omega \\ y(0) = y_{xx}(0) = y(L) = y_{xx}(L) = 0, & \text{on } (0, T) \times \Omega \\ y(0) = y^0, \quad y_t(0) = y^1, & \text{in } (0, L) \times \Omega, \end{cases} \quad (7.7)$$

where $\alpha = \alpha\,(\omega)$ is a random variable uniformly distributed in the interval $[-\varepsilon, \varepsilon]$, with $0 < \varepsilon < 1$.

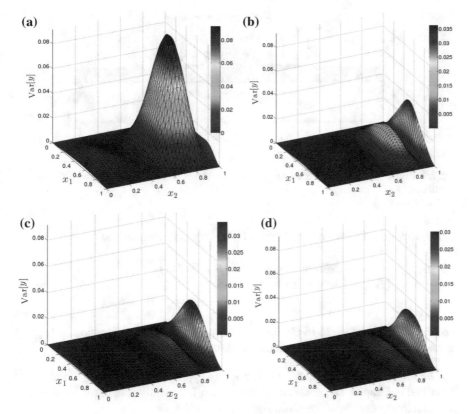

Fig. 7.5 Problem (P_{heat}). Variance of the state variable associated with the cost J_2 for $\gamma_2 = 1e - 8$, $\gamma_1 = 0$ (**a**), $\gamma_1 = 1$ (**b**), $\gamma_1 = 2$ (**c**), and $\gamma_1 = 3$ (**d**). From [11], reproduced with permission

The cost functional considered is:

$$J(u) := \frac{\delta_1}{2} \int_0^L \left(\int_\Omega y(T)\, d\mathbb{P}(\omega) \right)^2 dx + \frac{\delta_2}{2} \int_0^L \left(\int_\Omega y_t(T)\, d\mathbb{P}(\omega) \right)^2 dx$$

$$+ \frac{\beta_1}{2} \int_0^L \int_\Omega y^2(T)\, d\mathbb{P}(\omega) dx + \frac{\beta_2}{2} \int_0^L \int_\Omega y_t^2(T)\, d\mathbb{P}(\omega) dx \qquad (7.8)$$

$$+ \frac{\gamma}{2} \int_0^T \int_{\mathscr{O}} u^2\, dx dt,$$

where $\delta_1, \delta_2 > 0$ and $\beta_1, \beta_2, \gamma \geq 0$ are weighting parameters.

The rest of input parameters in (7.7) are: $L = 1$, $T = 0.5$, $\mathscr{O} = (0.2, 0.8)$ and $\left(y^0(x), y^1(x) \right) = (\sin(\pi x), 0)$. See Fig. 7.6 for the problem configuration.

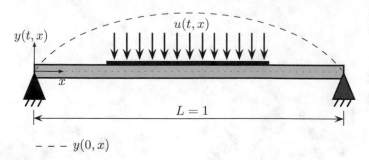

Fig. 7.6 Problem (P_{beam}). Geometry, initial and boundary conditions. From [11], reproduced with permission

Eventually, the robust optimal control problem is formulated as:

$$(P_{beam}) \begin{cases} \text{Minimize in } u \in L^2(\mathcal{O}): \quad J(u) \\ \text{subject to} \\ \qquad\qquad\qquad y = y(u) \text{ solves (7).} \end{cases}$$

Remark 7.1 For the shake of clarity and simplicity, the control function in (P_{beam}) is not of piezoelectric type, but a more academic control $u \in L^2(\mathcal{O})$.

The solution of (P_{beam}) may be numerically approximated by using the methods described in Chap. 4. The reader is referred to [10] for details. Here, we focus on the physical interpretation of the results obtained.

Figure 7.7 shows the computed optimal control $u(t, x)$ in the cases: (a) deterministic problem, where $\alpha = 0$ in (7.7) and the cost functional is

$$J_{det}(u) = \frac{1}{2}\|y(T)\|^2_{L^2(0,L)} + \frac{1}{2}\|y_t(T)\|^2_{L^2(0,L)} + \frac{\gamma}{2}\|u\|^2_{L^2(]0,T[\times\mathcal{O})},$$

and (b)–(c) solutions of (P_{beam}) for different values of the weighting parameters and for $\varepsilon = 0.5$, i.e., $\alpha(\omega)$ is uniformly distributed in the interval $[-0.5, 0.5]$. It is observed that the optimal control obtained minimizing only the mean of the state variable (Fig. 7.7b) is quite similar to the one obtained in the deterministic case (Fig. 7.7a). However, the optimal control including the second raw moment in the cost functional (Fig. 7.7c) presents a very different behavior.

To better understand the qualitative behavior of the control profiles, the way in which the beam equation propagates uncertainty is analyzed in Fig. 7.8. Precisely, let us denote by $y_{unc}(t, x, \omega)$ the solution of the uncontrolled ($u = 0$ in (7.7)) beam equation. It is easy to see that

$$\begin{aligned} y_{unc}(t, x, \omega) &= \cos\left(\sqrt{1 + \alpha(\omega)}\pi^2 t\right)\sin(\pi x) \\ &= \tfrac{1}{2}\sin\left[\pi\left(x + \pi t\sqrt{1 + \alpha(\omega)}\right)\right] + \tfrac{1}{2}\sin\left[\pi\left(x - \pi t\sqrt{1 + \alpha(\omega)}\right)\right]. \end{aligned}$$
$$(7.9)$$

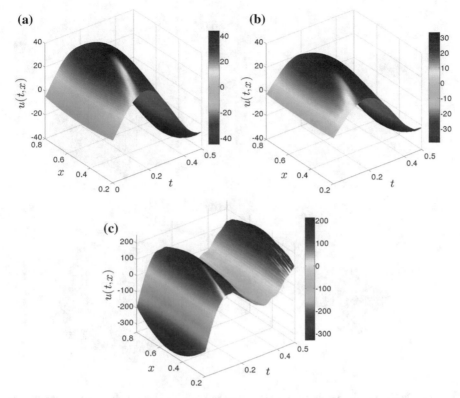

Fig. 7.7 Problem (P_{beam}). $\varepsilon = 0.5$ and $\gamma = 10^{-6}$. Optimal controls for: (**a**) deterministic problem, (**b**) $\delta_1 = \delta_2 = 1$, $\beta_1 = \beta_2 = 0$ and (**c**) $\delta_1 = \delta_2 = 1$, $\beta_1 = \beta_2 = 1$. From [10], reproduced with permission

From (7.9) one concludes: (i) the initial condition $y^0(x) = \sin(\pi x)$ propagates in time as a wave with random amplitude $\cos\left(\sqrt{1 + \alpha(\omega)}\pi^2 t\right)$, and (ii) each material point x propagates as the superposition of two waves travelling in opposite directions at a random speed $\pi\sqrt{1 + \alpha(\omega)}$. Moreover, from (7.9) one may explicitly compute the first two moments of $(y_{unc}, (y_t)_{unc})$, which are plotted in Fig. 7.8. In particular, by the mean value theorem for integrals,

$$\overline{y}_{unc}(t, x) = \frac{1}{2\varepsilon} \int_{-\varepsilon}^{\varepsilon} \cos\left(\sqrt{1 + \alpha}\pi^2 t\right) \sin(\pi x)\, d\alpha = \cos\left(\sqrt{1 + \alpha^\star}\pi^2 t\right) \sin(\pi x)$$

for some $-\varepsilon \leq \alpha^\star \leq \varepsilon$. Notice that $\alpha^\star$ depends on t. Thus, for ε small, $\overline{y}_{unc}$ is similar to the uncontrolled deterministic solution, where $\alpha = 0$ in (7.9). This fact explains the similarity between the controls plotted in Fig. 7.7a–b. However, the second order moments of $(y_{unc}, (y_t)_{unc})$, depicted in Fig. 7.8c–d, show a very different qualitative behavior. Precisely, one observes an additional oscillation and there is a change in amplitude. The control displayed in Fig. 7.7c is in accordance with this behavior of the second moment of the uncontrolled random solution.

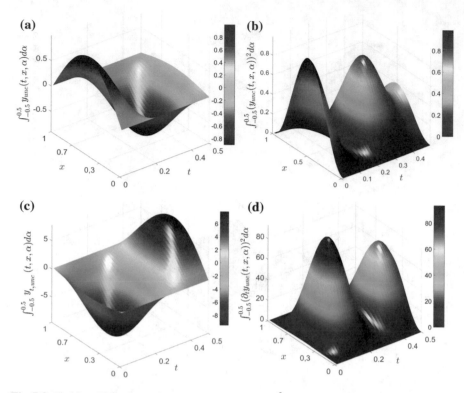

Fig. 7.8 Problem (P_{beam}). $\varepsilon = 0.5, \delta_1 = \delta_2 = 1, \gamma = 10^{-6}$. First column displays: (**a**) expectation of the uncontrolled solution $y_{unc}(t, x, \omega)$, and (**b**) expectation of $(y_t)_{unc}(t, x, \omega)$. Second column shows: (**c**) second moment of $y_{unc}(t, x, \omega)$, and (**d**) second moment of $(y_t)_{unc}(t, x, \omega)$. From [10], reproduced with permission

7.3 Concluding Remarks and Some Open Problems

As it has been noticed along this introductory text, a major computational challenge in the field of control under uncertainty is *breaking the curse of dimensionality*. Indeed, when the number of random variables, which appear in the uncertain parameters, is large, the number of degrees of freedom grows exponentially so that the problem may be computationally unaffordable. This issue is exacerbated, for instance, in:

- multiphysics and/or multiscale problems, where the solution of the underlying PDE is expensive, and in
- nonlinear PDEs, where numerical resolution methods typically require iteration.

A number of remedies are being proposed to overcome these difficulties. For instance, Analysis of Variance (ANOVA) based techniques, High-Dimensional Model Representation (HDMR) or a combination of them, aim at detecting the most influential random variables (and the interaction between them) in the mathematical model (see [6, 13] and the references therein). Other model order reduction methods, which are receiving an increasing interest in the last decades, are Reduced Basis (RB) methods, Proper-Orthogonal Decomposition (POD) methods [1, 4] and Active Subspaces methods [3], among others.

Despite these theoretical and numerical developments, the topic of control of random PDEs may be still considered to be in its infancy and further research is needed to better understand and more efficiently solve this type of problems.

Another class of control problems, which has not been considered in this text, is exact controllability of PDEs under uncertainty. As an illustration of this type of problems, let us consider again the controlled heat equation

$$
\begin{cases}
y_t - \mathrm{div}\,(a\nabla y) = 0, & \text{in } (0, T) \times D \times \Omega \\
a\nabla y \cdot n = 0, & \text{on } (0, T) \times \partial D_0 \times \Omega \\
a\nabla y \cdot n = \alpha\,(u - y), & \text{on } (0, T) \times \partial D_1 \times \Omega, \\
y\,(0) = y^0 & \text{in } D \times \Omega,
\end{cases}
\tag{7.10}
$$

where, for instance, the parameters $a = a\,(x, \omega)$, $\alpha = \alpha\,(x, \omega)$ and the initial condition $y^0 = y^0\,(x, \omega)$ are assumed to be uncertain, and $u \in L^2\,((0, T) \times \partial D_1)$ is the control function.

The concept of averaged exact control was introduced by Zuazua in [14]. When applied to system (7.10), the problem of averaged null controllability is formulated as: given a control time $T > 0$, find $u \in L^2\,((0, T) \times \partial D_1)$ such that the mean of the solution $y = y\,(u)$ to (7.10) satisfies

$$
\int_\Omega y\,(T, x, \omega)\, d\mathbb{P}\,(\omega) = 0, \quad x \in D.
\tag{7.11}
$$

At the theoretical level, some positive averaged controllability results have been obtained for the heat, wave and Schrödinger equations in [7–9]. In all these works uncertainty appears in the coefficient a and is represented either as a single random variable or as a small random perturbation. The general case of uncertainties modelled as random fields and/or located in other input parameters of the model such as geometry, location of controls or boundary conditions has not been addressed so far. It is then interesting to analyze if classical tools in deterministic exact controllability problems such as Ingham's inequalities or Carlerman estimates, among others, may be adapted to deal with this class of controllability problems.

The numerical approximation of averaged exact controllability problems is also a challenging problem. By using an optimal control approach, a first step in this direction has been given in [10, 11].

Related to problem (7.10)–(7.11) is the following robust version:

$$
(P_R) \begin{cases}
\text{Minimize in } u \in L^2\,((0, T) \times \partial D_1) : \|\mathrm{Var}\,(y(T))\|^2_{L^2(D)} \\
\text{subject to} \\
\qquad\qquad (y, u) \text{ solve } (7.10)\text{–}(7.11).
\end{cases}
\tag{7.12}
$$

Up to the best knowledge of the authors, problem (P_R) is open.

References

1. Chen, P., Quarteroni, A.: A new algorithm for high-dimensional uncertainty quantification based on dimension-adaptive sparse grid approximation and reduced basis methods. J. Comput. Phys. **298**, 176–193 (2015)
2. Chiba, R.: Stochastic heat conduction of a functionally graded annular disc with spatially random heat transfer coefficients. Appl. Math. Model. **33**(1), 507–523 (2009)
3. Constantine, P.G.: Active subspaces. Emerging ideas for dimension reduction in parameter studies, vol. 2. SIAM Spotlights, Philadelphia, PA, 2015
4. Hesthaven, J.S., Rozza, G., Stamm, B.: Certified reduced basis methods for parametrized partial differential equations. SpringerBriefs in Mathematics. BCAM SpringerBriefs (2016)
5. Kuznetsov, A.V.: Stochastic modeling of heating of a one-dimensional porous slab by a flow of incompressible fluid. Acta Mech. **114**, 39–50 (1996)
6. Labovsky, A., Gunzburger, M.: An efficient and accurate method for the identification of the most influential random parameters appearing in the input data for PDEs. SIAM/ASA J. Uncertain. Quantif. **2**(1), 82–105 (2014)
7. Lazar, M., Zuazua, E.: Averaged control and observation of parameter-depending wave equations. C. R. Acad. Sci. Paris. Ser. I **352**, 497–502 (2014)
8. Lohéac, J., Zuazua, E.: Averaged controllability of parameter dependent wave equations. J. Differ. Equ. **262**(3), 1540–1574 (2017)
9. Lü, Q., Zuazua, E.: Averaged controllability for random evolution partial differential equations. J. Math. Pures Appl. **105**(3), 367–414 (2016)
10. Marín, F.J., Martínez-Frutos, J., Periago, F.: Robust averaged control of vibrations for the Bernoulli-Euler beam equation. J. Optim. Theory Appl. **174**(2), 428–454 (2017)
11. Martínez-Frutos, J., Kessler, M., Münch, A., Periago, F.: Robust optimal Robin boundary control for the transient heat equation with random input data. Internat. J. Numer. Methods Eng. **108**(2), 116–135 (2016)
12. Nobile, F., Tempone, R.: Analysis and implementation issues for the numerical approximation of parabolic equations with random coefficients. Internat. J. Numer. Methods Eng. **80**(6–7), 979–1006 (2009)
13. Smith, R. C.: Uncertainty Quantification. Theory, Implementation and Applications. Comput. Sci. Eng. **12** (2014)
14. Zuazua, E.: Averaged control. Automatica **50**, 3077–3087 (2014)

Index

© The Author(s), under exclusive license to Springer Nature Switzerland AG 2018
J. Martínez-Frutos and F. Periago Esparza, *Optimal Control of PDEs under Uncertainty*,
SpringerBriefs in Mathematics, https://doi.org/10.1007/978-3-319-98210-6

Printed in the United States
By Bookmasters